Biological Magnetic Materials and Applications

Tadashi Matsunaga • Tsuyoshi Tanaka
David Kisailus
Editors

Biological Magnetic Materials and Applications

Springer

Editors
Tadashi Matsunaga
Tokyo University of Agriculture and Technology
Fuchu, Tokyo, Japan

David Kisailus
Department of Chemical and Environmental Engineering
University of California, Riverside
Riverside, CA, USA

Tsuyoshi Tanaka
Division of Biotechnology and Life Science, Institute of Engineering
Tokyo University of Agriculture and Technology
Koganei, Tokyo, Japan

ISBN 978-981-10-8068-5 ISBN 978-981-10-8069-2 (eBook)
https://doi.org/10.1007/978-981-10-8069-2

Library of Congress Control Number: 2018946099

Printed on acid-free paper

This Springer imprint is published by the registered company Springer Nature Singapore Pte Ltd.
The registered company address is: 152 Beach Road, #21-01/04 Gateway East, Singapore 189721, Singapore

Preface

Over hundreds of millions of years, biological systems have developed strategies to synthesize and assemble nanomaterials with a precision yet to be achieved by artificial synthetic methods. Using abundant elements in earth's crust, these biological materials are formed under mild conditions, at near neutral pH and ambient temperatures. These natural synthetic pathways have been developed over many millennia, providing potential alternative strategies to synthesize next-generation nanomaterials. One example of utilization of biological synthetic pathways for controlled nanoarchitectures comes from iron oxide magnetic materials. The first finding of magnetic materials in organisms was from the magnetic teeth of mollusks. The presence of magnetite (Fe_3O_4)-based nanorods in the teeth provides excellent mechanical properties and highlights their utility to harvest microalgae growing on and within rocky substrates. In addition, magnetotactic bacteria produce chains of magnetic iron oxide or iron sulfide nanoparticles. They use the magnetic nanoparticles as biological geomagnetic sensors that help them to navigate and survive in their habitats on earth. Based on recent fundamental studies on the formation mechanisms of magnetic materials in these organisms, novel structures, biological molecules, and formation principles have been revealed. These findings are now attracting chemists, materials scientists, and engineers to translate knowledge gleaned from these systems toward the development of new advanced magnetic materials for future technologies. In addition, new functional materials are being developed by utilizing genetically modified organisms and the underlying biotemplating proteins found in biologically synthesized magnetic materials.

Magnetic materials are widely used for numerous applications in the field of electronics, mechanical engineering, and biotechnology. Magnetic nanomaterials are important for ultrahigh density magnetic recording media. They are also used for biomedical processes such as magnetic resonance imaging, cell separation, environmental inspections, drug delivery, and hyperthermia. It is also shown that magnetic iron oxide particles have catalytic utility in polymer syntheses. Therefore, this book focuses on iron-based biological magnetic materials and

1. Leverages a diversity of multidisciplinary expertise encompassing cellular and molecular biology, paleobiology, materials science and engineering, bio- and nanotechnology, and medical sciences
2. Provides a wide scope – balancing basic science with large-scale engineering (cradle to grave strategies), yet includes sufficient detail for scientific accumulation

This book consists of three major sections: (I) formation mechanisms of biological magnetic materials, (II) biological templating of magnetic materials for medical and device applications, and (III) bio-inspired magnetic materials for nanotechnology-based applications. The approach is from multiple perspectives: (i) organisms which utilize biologically derived magnetic materials and their function; (ii) formation mechanisms of biologically derived magnetic materials; (iii) how these biological formation routes yield various phase and morphologically derived magnetic materials; (iv) resultant properties (i.e., magnetic, structural). From this, the book describes the utilization of bio-inspiration to develop new materials for applications ranging from machine to bio-industries. Furthermore, we want to highlight recent strategies for developing scalable technologies that encompass protein displays, drug delivery, biophysical separations, and medical diagnostics as well as future insights toward next-generation applications. This book provides insights for scientists who want to access multidisciplinary fields to overcome current barriers and develop a multifaceted expertise in bionanotechnology.

Finally, the editors wish to thank all the authors of the chapters, coordination from the publisher, and the many people who worked behind the scenes in making this book a reality.

Fuchu, Tokyo — Tadashi Matsunaga
Koganei, Tokyo — Tsuyoshi Tanaka
Riverside, CA, USA — David Kisailus
July 2018

Contents

Part I
Formation Mechanisms of Biological Magnetic Materials

Chapter 1
Structure and Function of Aligned Magnetic Crystals in Magnetotactic Bacteria

Azuma Taoka and Yoshihiro Fukumori

Abstract Magnetotactic bacteria swim along the magnetic field using a unique bacterial organelle termed the magnetosome. Magnetosomes are comprised of nano-sized crystals of magnetite or greigite enclosed within lipid bilayer membrane vesicles and specifically associated proteins. The integration of magnetosomes in motility allows magnetotactic bacteria to orient themselves to find a favorable microaerobic habitat. Most of the magnetosome-associated proteins are encoded in gene clusters within a "magnetosome island." Additionally, these proteins are highly conserved and essential for the synthesis and maintenance of magnetosomes. In this chapter, we will briefly introduce general insight into magnetotactic bacteria, and then, we will present recent research progress on magnetosome structure and function.

1.1 Introduction

Magnetotactic bacteria are widespread aquatic eubacteria that utilize organelles known as magnetosomes to navigate along Earth's magnetic field (reviewed in Bazylinski and Frankel 2004; Faivre and Schüler 2008; Komeili 2012; Lin et al. 2013; Schüler 2008). The unique properties of these bacteria regarding magnetic sensing, magnetosome formation, and biomineralization have garnered much multidisciplinary interests from fields such as microbiology, geology, chemistry, physics, mineralogy, astrobiology, synthetic biology, bioengineering, materials science, chemical engineering, medicine, and bionanotechnology. In this first section, we will provide a general overview of magnetotactic bacteria.

A. Taoka (✉) · Y. Fukumori
Faculty of Biological Science and Technology, Institute of Science and Engineering, Kanazawa University, Kanazawa, Japan
e-mail: aztaoka@staff.kanazawa-u.ac.jp

T. Matsunaga et al. (eds.), *Biological Magnetic Materials and Applications*,
https://doi.org/10.1007/978-981-10-8069-2_1

1.1.1 Discovery of Magnetotactic Bacteria

Two scientists, Richard Blakemore and Salvatore Bellini, independently and serendipitously discovered magnetotactic bacteria (Frankel 2009). Magnetotactic bacteria were first demonstrated by Salvatore Bellini, a medical doctor at the University of Pavia in Italy. He found magnetotactic bacteria in freshwater samples collected from Pavia and studied them from the late 1950s until 1963. His work was published in 1963 but was not widely disseminated. However, his findings have recently come to light, and his papers were translated to English and subsequently published in an international journal (Bellini 2009a, b). Richard Blakemore rediscovered magnetotactic bacteria in 1974 (Blakemore 1975). While observing sediment samples collected near Woods Hole, Massachusetts to study microbes enriched in sulfide mud, he found a group of highly motile bacteria that swam continuously in the same geographic direction (Blakemore 1982). These bacteria altered their swimming direction by placing a bar magnet near the microscope slide, and he deduced that these bacteria swam along the magnetic field. Blakemore demonstrated that these bacteria contained chains of iron-rich magnetic crystals in their cells, thus providing an explanation for their ability to respond to magnetic fields (Blakemore 1975). He termed such cell motility directed by a magnetic field as "magnetotaxis" (Blakemore 1975).

1.1.2 Isolation of Magnetotactic Bacteria

Magnetotactic bacteria are ubiquitous, widely distributed in a diverse array of aquatic environments such as fresh, marine, brackish, and hypersaline water or in chemically stratified water columns (Lefèvre and Bazylinski 2013). These bacteria have also been found on all continents (Lin et al. 2013; Lefèvre and Wu 2013). Although most magnetotactic bacteria dwell in aquatic habitats at ambient temperature and neutral pH, some species are extremophiles. Lefèvre et al. recently reported moderately thermophilic magnetotactic bacteria in hot springs (Lefèvre et al. 2010) and obligate magnetotactic alkaliphiles magnetotactic bacteria living in alkaline lakes (Lefèvre et al. 2011b).

Magnetotactic bacteria are either microaerobes or anaerobes and are found at the oxic-anoxic interface or anoxic region of aquatic habitats, respectively, in sediments or at the bottom of water column. Here we briefly mention a simple method to isolate magnetotactic bacteria from freshwater sediments. Sediment (0–3 cm deep) and surface water are collected from near the edge of a shallow freshwater pond using a scooper and placed into tightly capped 0.5 L glass bottle. The magnetotactic bacteria are concentrated by attaching a bar magnet to the outside of the bottle just above the sediment/water interface. North-seeking magnetotactic bacteria (swimming toward the magnetic S-pole) dominate in the Northern Hemisphere, while south-seeking magnetotactic bacteria (swimming toward the magnetic N-pole) dominate in the Southern Hemisphere. Therefore, for example, the S pole of the bar magnet is attached to the glass bottle in case of observation in the North Hemisphere. The bottle is kept in the dark for up to 3 h to allow magnetotactic cells to swim toward the magnet. A pipet is then used to collect magnetotactic bacteria. A drop of the magnetotactic bacteria is placed onto a slide, and the S-pole of bar magnet is placed on the microscope stage near the drop. The magnetotactic bacteria will swim to the edge of the drop near the magnet (Fig. 1.1a). When the bar magnet is rotated 180°, the cells will rotate and swim away from the magnet. Figure 1.1b shows examples of magnetotactic bacteria that were collected from a freshwater pond in Kanazawa, Japan. Morphologically diverse magnetotactic bacteria were observed from this collection, with chains of magnetosomes to help the bacteria move along the magnetic field.

1.1.3 Magnetotactic Bacteria Diversity

Magnetotactic bacteria are a physiologically, morphologically, and phylogenetically diverse group of Gram-negative bacteria. Many bacteria from this group have been isolated from various aquatic environments; however, the surface of magnetotactic bacteria diversity has barely been scratched due to difficulty in obtaining axenic cultures. Despite their ubiquitous presence, magnetotactic bacteria are fastidious and grow slowly. Recently, there has been much improvement in isolation and culturing methods, and employing current genetic techniques, such as metagenomics and single-cell genome analysis, has substantially provided insight about the diversity of magnetotactic bacteria. Known magnetotactic bacteria belong to different *Proteobacteria* phylum subgroups, including at least *alpha-*, *gamma-*, and *delta-*subgroups; the *Nitrospirae* phylum; and the OP3 of the PVC (*Planctomycetes*, *Verrucomicrobia*, and *Chlamydiae*) superphylum (reviewed in Lefèvre and Bazylinski 2013) (Table 1.1). Thus, magnetotactic bacteria do not form a phylogenetically coherent group.

With over 40 years of sustained efforts to date, more than 20 species of axenic cultures have been obtained. To note, species belonging to *Alphaproteobacteria*, e.g., *Magnetospirillum magnetotacticum* MS-1, *M. magneticum* AMB-1, *M. gryphiswaldense* MSR-1, *Magnetococcus marinus* MC-1, and *Magnetovibrio blakemorei* MV-1; *Gammaproteobacteria*, e.g., strain BW-2 and SS-5; and *Deltaproteobacteria*, e.g., *Desulfovivbrio magneticus* RS-1, have been cultivated (Table 1.1). Specifically, *M. magneticum* AMB-1 and *M. gryphiswaldense* MSR-1 are used as model species for the genetic and biochemical dissection of magnetosome formation and function. However, species belonging to *Nitrospirae*, and the PVC superphylum have yet to be cultured.

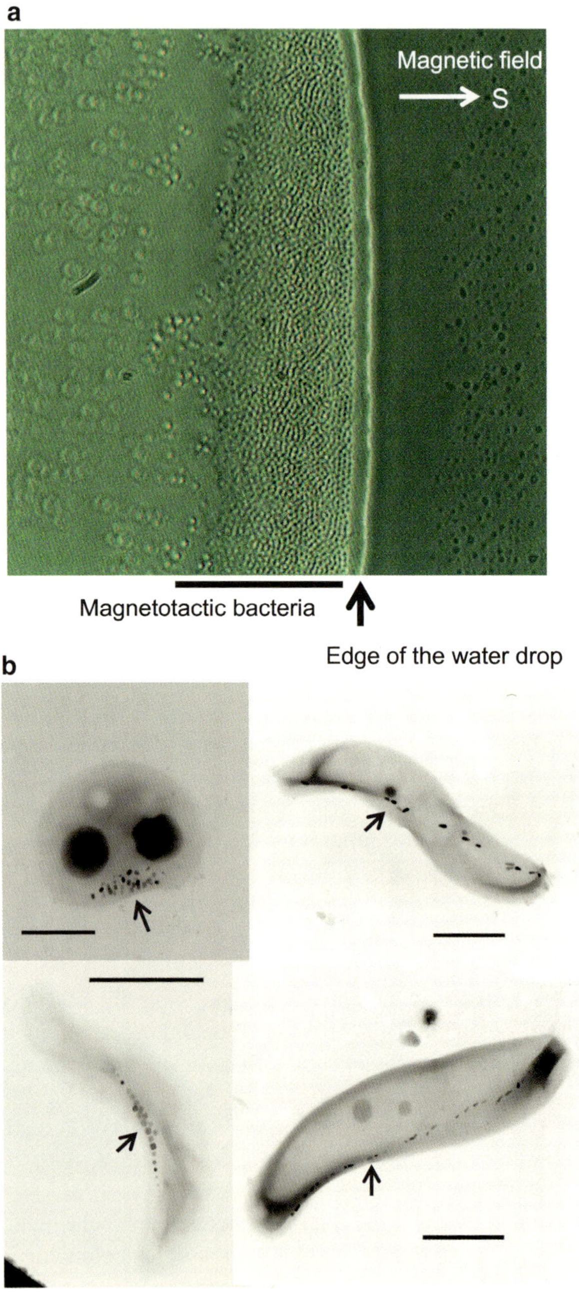

Fig. 1.1 Isolation of magnetotactic bacteria. (**a**) Phase contrast microscopic image of magnetotactic bacteria in a water drop. The S-pole of the bar magnet was placed on the right side of the image. Magnetotactic bacteria accumulated at the edge of the water drop. (**b**) Transmission electron microscopic (TEM) images of isolated magnetotactic bacteria from a freshwater pond in Kanazawa, Japan. Note that all of the cells possess chains of opaque structures, magnetosomes (arrows), to facilitate navigations along the magnetic field. Bar: 1 μm

Table 1.1 Isolated and characterized species of magnetotactic bacteria

Phyla (classes)	Genus and species[a]	References
Alphaproteobacteria	*Magnetospirillum magneticum* AMB-1	Matsunaga et al. (1991)
	Magnetospirillum sp. WM-1	Li et al. (2007)
	Magnetospirillum magnetotacticum MS-1	Blakemore et al. (1979)
	Magnetospirillum gryphiswaldense MSR-1	Schleifer et al. (1991)
	Magnetospirillum aberrantis	Gorlenko et al. (2011)
	Magnetospirillum bellicus	Thrash et al. (2010)
	Magnetospirillum sp. XM-1	Wang et al. (2015)
	Magnetospirillum sp. VITRJS1	Revathy et al. (2016)
	Magnetococcus marinus MC-1	Meldrum et al. (1993b)
	Strain MO-1	Lefèvre et al. (2009)
	Magnetofaba australis IT-1	Zhang et al. (2012)
	Magnetovibrio blakemorei MV-1	Bazylinski et al. (1988)
	Strain MV-2	Meldrum et al. (1993a)
	Magnetospira thiophila MMS-1	Williams et al. (2012)
	Magnetospira sp. QH-2	Zhu et al. (2010)
Gammaproteobacteria	Strain BW-2, and SS-5	Lefèvre et al. (2012)
	Strain GRS-1	Taoka et al. (2014)
Deltaproteobacteria	*Ca.* Magnetoglobus multicellularis	Abreu et al. (2007)
	Ca. Magnetomorum litorale	Wenter et al. (2009)
	Ca. Magnetananas tsingtaoensis	Zhou et al. (2012)
	Desulfovibrio magneticus RS-1	Sakaguchi et al. (1993)
	Strain ZZ-1, ML-1, AV-1	Lefèvre et al. (2011b)
	Desulfamplus magnetovallemortis BW-1, SS-2	Lefèvre et al. (2011c)
Nitrospirae	*Ca.* Magnetobacterium bavaricum	Petersen et al. (1989)
	Ca. Magnetobacterium bremense MHB-1	Flies et al. (2005)
	Ca. Magnetobacterium casensis	Lin et al. (2014)
	Ca. Thermomagnetovibrio paiutensis HSMV-1	Lefèvre et al. (2010)
	Ca. Magnetoovum mohavensis LO-1	Lefèvre et al. (2011a)
	Ca. Magnetoovum chiemensis CS-04	Kolinko et al. (2013)
	Strain MWB-1	Lin et al. (2012)
PVC superphylum	*Ca.* Omnitrophus magneticus SKK-01	Kolinko et al. (2012)

[a]*Ca. Candidatus*

1.1.4 Magnetotaxis

Magnetotaxis is thought to facilitate cell location and maintenance cells within the microaerobic zones on the bottom of chemically stratified natural waters or in sediments (Frankel et al. 1997). The intracellular magnetic dipoles of magnetosomes enable bacteria to orient themselves to swim parallel to the geomagnetic field (Fig. 1.2). Due to the inclination of the Earth's magnetic field (white dash arrows, Fig. 1.2), magnetotactic bacteria can swim unidirectionally along the vertical axis in aquatic habitats. Interestingly, north-seeking bacteria occur predominantly in the Northern Hemisphere, while south-seeking bacteria are predominantly in the

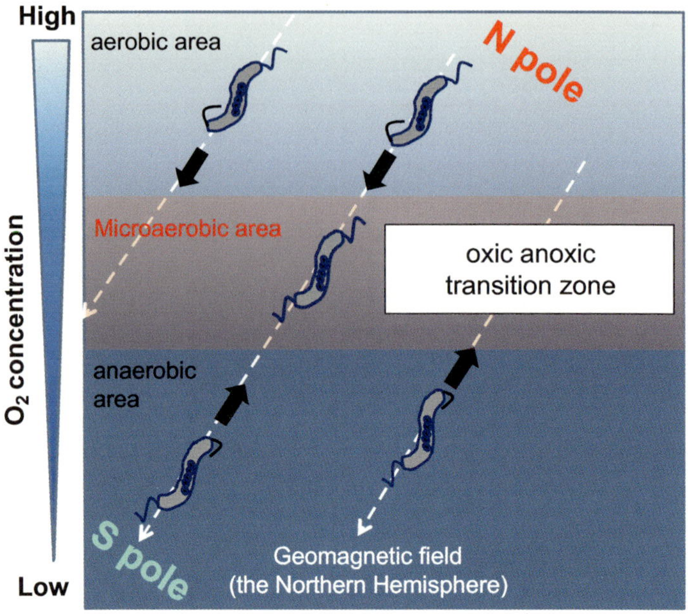

Fig. 1.2 Magneto-aerotaxis model

Southern Hemisphere. In addition, the abundance of north- and south-seeking bacteria is distributed equally at the equator (Blakemore 1975, 1982). As a consequence, both of Northern Hemisphere magnetotactic bacteria and Southern Hemisphere magnetotactic bacteria swim to regions of low oxygen. Thus, magnetotactic bacteria swim along the geomagnetic field by using magnetotaxis. These bacteria maintain their position within the microaerobic environment via aerotaxis. This behavior is collectively termed "magneto-aerotaxis" (Frankel et al. 1997).

1.2 Magnetosome Structure and Protein Localization

Magnetosomes contain membrane-enclosed regular-sized magnetite (Fe_3O_4) or greigite (Fe_3S_4) crystals. Individual magnetosome particles are aligned into one or multiple chains and localize along the long axis of the cell, thereby facilitating cellular orientation along the geomagnetic field and the search for a microaerobic environment. Magnetosomes consist of the Mam and Mms proteins, which are specifically localized. These proteins function in magnetosome formation and are critical for precise control of the biomineralization of magnetite crystals and magnetic reception. In this section, we present our recent studies dissecting the roles of magnetosome-associated proteins MamA and MamK in maintenance of magnetosome structure.

1.2.1 Magnetosome-Associated Proteins

Figure 1.3a shows magnetosomes isolated from *M. magneticum* AMB-1, and Fig. 1.3b shows the SDS-PAGE gel profile of magnetosome-associated proteins. In 2004, Grünberg et al. identified key proteins to magnetosome function via proteomics and found that magnetosome-associated proteins are encoded in a specific genomic region called the magnetosome island (MAI) (Fig. 1.4) (Schübbe et al. 2003; Ullrich et al. 2005; Jogler et al. 2009; Fukuda et al. 2006). All magnetotactic bacterial genomes identified thus far have MAIs. The MAI of *Magnetospirillum* species encodes magnetosome-associated proteins in four operons, *mamGFDC*, *mms6, mamAB*, and *mamXY* (Fig. 1.4) (Matsunaga et al. 2005). Magnetosome-associated proteins comprise various kinds of proteins, such as TPR protein, transporter, PDZ domain-containing proteins, cytoskeletal proteins, heme proteins, serine proteases, and hypothetical proteins. These proteins show no or very low homology to known proteins and thus specific to magnetosome function, such as magnetosome biogenesis (Richter et al. 2007).

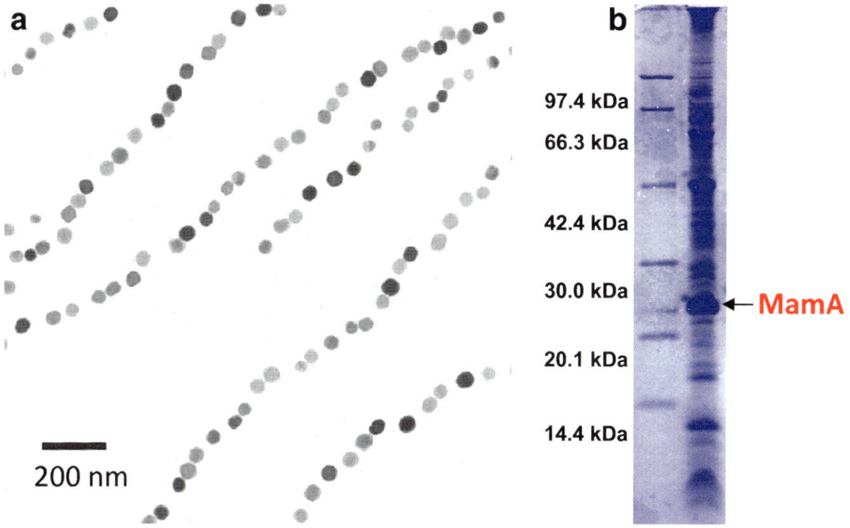

Fig. 1.3 Isolated magnetosomes. (**a**) TEM image of isolated magnetosomes. (**b**) SDS-PAGE gel profile of extracted proteins from isolated magnetosomes. MamA (arrow) is one of the most abundant proteins in the magnetosome

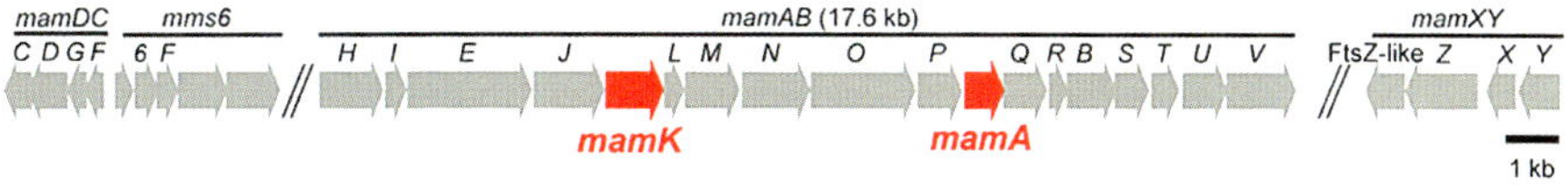

Fig. 1.4 The genomic organization of the *M. magneticum* AMB-1 magnetosome island (MAI) consists of four operons: *mamGFDC*, *mms6*, *mamAB*, and *mamXY*

1.2.2 Tetratricopeptide Repeat (TPR) Protein MamA

MamA (Mam22) was the first identified magnetosome-associated protein (Okuda et al. 1996; Okuda and Fukumori 2001). MamA is also one of the most abundant proteins and conserved in MAIs of all known magnetotactic bacteria (Lefèvre et al. 2013). Taken together, MamA may have an essential role in the magnetosome. The primary structure of MamA consists of five TPR motifs and one putative TPR motif (Okuda and Fukumori 2001). A single TPR motif adopts a helix-turn-helix fold, while several adjacent TPR motifs create antiparallel α-helices that form a superhelix structure. This yields a pair of concave and convex curved surfaces that display binding sites capable of forming a multiprotein complex (Zeytuni and Zarivach 2012). Yamamoto et al. and Zeytuni et al. determined that MamA proteins interact with themselves to form an oligomeric complex (Yamamoto et al. 2010; Zeytuni et al. 2011). According to atomic force microscopic (AFM) observation, MamA forms spherical-shaped oligomers, and the size of the oligomer ranged from 4.5 to 6.5 nm in height and 14 to 20 nm in diameter (Yamamoto et al. 2010). Recently, X-ray crystal structures of MamA from *M. magneticum* AMB-1 (Zeytuni et al. 2011), *M. gryphiswaldense* MSR-1 (Zeytuni et al. 2011), *Ca.* Magnetobacterium bavaricum Candidatus Magnetobacterium bavaricum (Zeytuni et al. 2012), and *Desulfovibrio magneticus* RS-1 (Zeytuni et al. 2015) were determined. Zeytuni et al. proposed that MamA contains at least three protein binding sites: a putative TPR binding site, a concave binding site, and a convex binding site (Zeytuni et al. 2011). MamA structural insight suggested that such protein-protein interactions are critical to its function in the magnetosome.

1.2.3 MamA and the Magnetosome Matrix

We observed isolated magnetosomes chains via transmission electron microscopy (TEM) and found that magnetosome vesicles are covered by a proteinaceous layer known as the magnetosome matrix (Taoka et al. 2006). We investigated MamA localization in magnetosomes using immunogold staining. Interestingly, MamA proteins localized to the magnetosome matrix. To further understand magnetosome structure and MamA localization, the use of AFM is employed. AFM is used to visualize organic samples under physiological conditions (Dufrêne 2008; Engel and Muller 2000). In AFM, the surface profile of the sample is imaged by detecting the interaction between the sample and the AFM stylus during the raster scanning of the sample. Figure 1.5a, b show AFM images of isolated magnetosomes. Based on AFM measurements, a magnetite crystal is surrounded by organic material with a ~7 nm thickness (Yamamoto et al. 2010). Upon labeling magnetosomes with anti-MamA antibodies and visualizing via AFM, we observed dense packing of anti-MamA antibodies on the magnetosomes (Fig. 1.5c–f), which indicated that MamA was located at the outermost layer of the magnetosomes to form the magnetosome

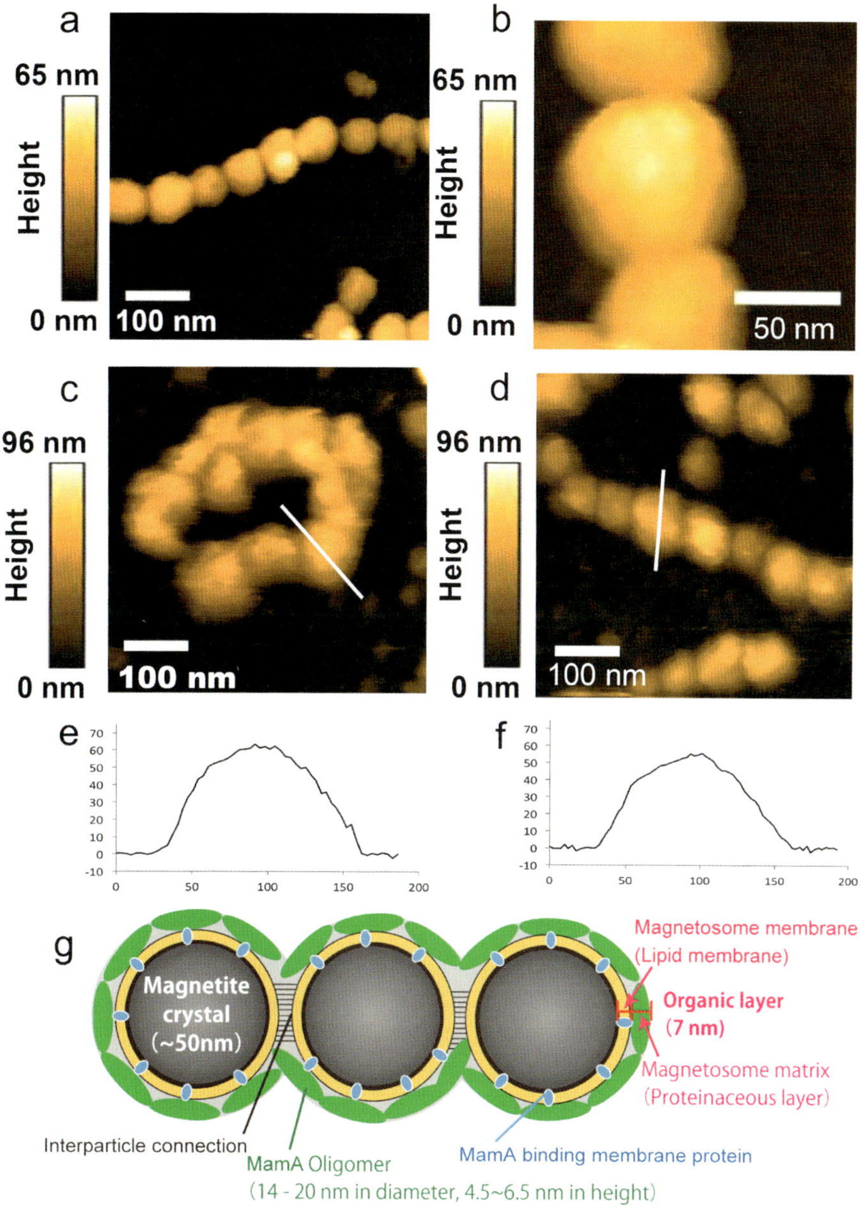

Fig. 1.5 AFM observations of isolated magnetosome. (**a**, **b**) AFM images of magnetosomes chains. (**c**, **d**) AFM images of immunolabeled magnetosomes. Magnetosomes labeled with (**c**) anti-MamA antibodies and (**d**) preimmune serum. (**e**, **f**) Surface profiles along the lines indicated in panels c and d. (**g**) Model of magnetosome organization

matrix (Fig. 1.5g). Moreover, we used AFM to observe chains of magnetosomes with and without MamA, and we proposed that MamA is anchored to the magnetosome membrane and may stabilize the structure of magnetosome chain (Yamamoto et al. 2010). Figure 1.5g shows a proposed model of magnetosome organization. In this model, MamA oligomers bind to the surface of magnetosome vesicles and form a proteinaceous layer to stabilize magnetosome chain arrangement. Recently, we reported that the magnetosome membrane protein Mms6 binds to MamA oligomers in vitro, suggesting that MamA and Mms6 interact each other in the magnetosome (Nguyen et al. 2016).

1.2.4 Cytoskeletal Filaments Associated with Magnetosome Chains

In 2006, two research groups independently confirmed the existence of cytoskeletal structures associated with magnetosome chains using cryo-electron tomography. Scheffel et al. found a network of cytoskeleton-like filamentous structures extending up to the cell pole of *M. gryphiswaldense* MSR-1 (Scheffel et al. 2006). The magnetosome vesicles were arranged along the filamentous structure located close to the cytoplasmic membrane. Komeili et al. also found networks of filaments running parallel to individual *M. magneticum* AMB-1 magnetosomes (Komeili et al. 2006). Furthermore, based on a *mamK* deletion mutant, filaments appeared to be comprised of MamK, a bacterial homolog of actin (Komeili et al. 2006).

MamK is found only in magnetotactic bacteria, unlike other well-known bacterial actin-like proteins MreB and ParM (Shaevitz and Gitai 2010). MreB is generally conserved in rod- and spiral-shaped bacteria and associated with most spatially organized cellular processes, including cell morphogenesis, cellular polarity, protein localization, cell division, cell differentiation, and chromosome segregation. ParM is a plasmid-encoded actin-like protein and functions in the plasmid segregation process. MamK, MreB, and ParM are predicted to form phylogenetically and functionally distinct groups of prokaryotic actin-like proteins (Derman et al. 2009). The amino acid sequence similarity is >50% among the MamK proteins of different magnetotactic bacteria species. On the other hand, amino acid identity between *M. magnetotacticum* MS-1 MamK and MreB is 24% and 17% between *M. magnetotacticum* MS-1 MamK and human actin. The actin protein superfamily has five conserved sequence motifs that form the ATP-binding pocket and the interdomain hinge region (Bork et al. 1992). MamK proteins also have five conserved motifs, suggesting that MamK has biochemical properties similar to those of other actin homologs (Fig. 1.6). MamK is encoded by the *mamAB* operon of MAI (Fig. 1.4) and is conserved in most identified magnetotactic bacterial genomes. The *mamAB* operon is essential for magnetosome biosynthesis (Lohße et al. 2011; Murat et al. 2010), suggesting that MamK is important for magnetosome organization.

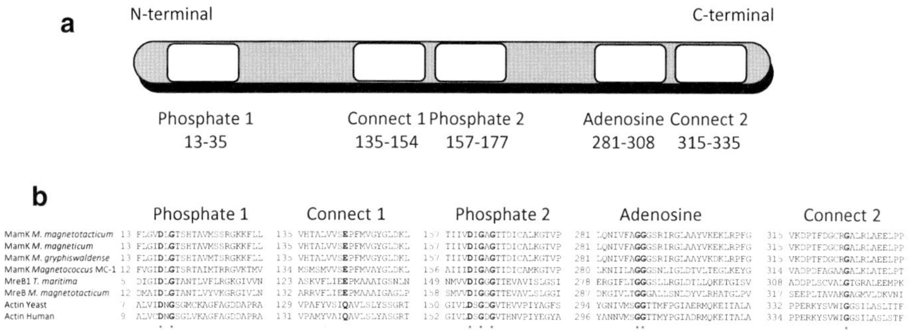

Fig. 1.6 Conserved motifs within the actin superfamily. (**a**) Schematic drawing of the five motifs in MamK that exhibit similarity to conserved motifs in proteins with an actin fold. Numbers indicated are amino acids in *M. magnetotacticum* MS-1 MamK. (**b**) Sequence alignments of the five conserved MamK motifs with MreB and eukaryotic actin proteins. Conserved amino acids are indicated in bold

1.2.5 MamK Localization

We compared the localization of MamK and MreB in early stationary phase *M. magnetotacticum* MS-1 cells to further characterize this cytoskeletal protein. The intracellular localization of MamK in *M. magnetotacticum* MS-1 was examined by both immunofluorescence and immunoelectron microscopy. Immunofluorescence microscopy with an anti-MamK antibody revealed a linear distribution of MamK from pole to pole along the vertical axis of the cells (Fig. 1.7a, b). In contrast, immunofluorescence microscopy using an anti-MreB antibody showed that MreB formed either spiral filamentous structures (Fig. 1.7c) or was located in the septa as paired dots (Fig. 1.7d). This distribution of MreB is consistent with that of other bacteria, such as *Bacillus subtilis* (Jones et al. 2001) and *E. coli* (Kruse et al. 2003). The results demonstrated that two kinds of actin-like proteins assemble as different filamentous structures in magnetotactic bacteria. Immunogold labeling of ultrathin sections confirmed the locality of MamK as shown in Fig. 1.7e.

To assess whether MamK filament is comprised of a single or multi-arranged protofilaments, the cellular abundance of the MamK protein was estimated by quantitative immunoblotting. The amount of MamK was estimated to be 26,000 ± 6000 ($n = 8$) molecules per cell in early stationary phase. Because of amino acid similarity and comparable molecular mass of MamK with MreB and ParM, the longitudinal monomer spacing of the MamK protein may be approximately that of MreB (51 Å) (van den Ent et al. 2001) and ParM (49 Å) (van den Ent et al. 2002). Thus, enough MamK is most likely available to accommodate a double helical filament of 14–22 times the length of the cell (mean length: 3.6 μm). These findings strongly suggest that MamK filaments exist as a bundle of multiple protofilaments or as a network structure in the cell. Additionally, in our analyses, MreB was estimated to be 5000 ± 1000 ($n = 6$) molecules per *M. magnetotacticum* cell. Thus, the difference

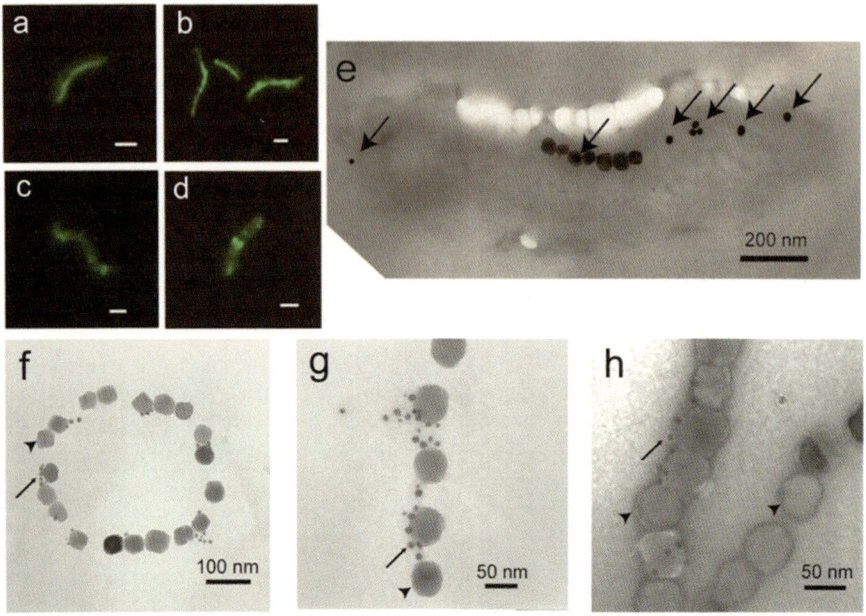

Fig. 1.7 Localization of MamK. (**a**, **b**) Immunofluorescence images using the anti-MamK antibody. (**c**, **d**) Immunofluorescence images using the anti-MreB antibody. (**e**) Immunogold labeling of an ultrathin section using anti-MamK indicating that MamK is distributed from pole to pole along the long axis of the cell (arrows indicate 15 nm gold particles). (**f**–**h**) Immunoelectron microscopic images of purified magnetosomes using anti-MamK antibodies that indicates the 5 nm gold particles (arrows) are distributed in a line along the surfaces of several magnetite particles (arrowheads). (**h**) Immunoelectron microscopic image of magnetosomes with negative staining (4% uranyl acetate). Scale bars (**a**–**d**): 1 μm

in the estimated values of MamK and MreB molecules in a single cell suggested that their expression is independently regulated.

To elucidate the localization of MamK in the magnetosome chain, we performed immunogold staining with isolated magnetosomes from *M. magnetotacticum* MS-1. Immunoelectron microscopy showed that the 5 nm gold particles were distributed in a line along the surfaces of several magnetite particles and appeared to remain on one side of the magnetosome chain with magnetite particles (Fig. 1.7f–h). These linear distributions of the gold particles suggested that MamK is filamentous.

1.2.6 MamK Polymerization

Next, we characterized MamK protein from *M. magnetotacticum* MS-1 in vitro (Taoka et al. 2007). The recombinant MamK protein was purified from *E. coli* and was added at a final concentration of 10 μM in a total volume of 30 μl polymerization buffer, which contained 100 mM Tris-HCl (pH 7.0), 14 mM $MgCl_2$, 100 mM

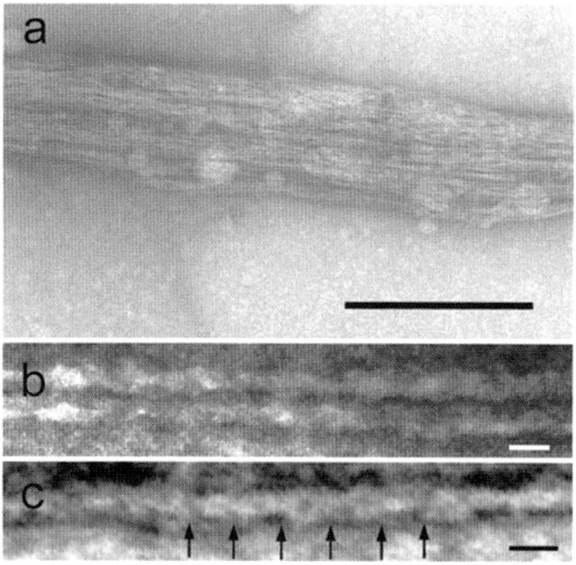

Fig. 1.8 TEM images of in vitro polymerized MamK filaments. (**a**) Low magnification of a MamK filamentous bundle. (**b**) Two filaments are indicated in the bundle. (**c**) Scanning TEM images of in vitro polymerized MamK filaments. A fine filament in the MamK filamentous bundle demonstrates that the filament is made up of two smaller helical filaments. Crossovers of the two helical filaments are indicated by the *arrows*. The mean crossover distance is 23 nm. The helical filament is approximately 6 nm wide (Scale bars, **a** 200 nm; **b** 20 nm, **c** 20 nm)

NaCl, and 30 mM KCl. The mixture was then centrifuged at 150,000 × *g* at 4 °C for 1 h to remove aggregated MamK. After the addition of 2 mM of the non-hydrolyzable ATP analog ATP-γ-S, the mixture was incubated at 30 °C for 5 min. The sample was then loaded on Formvar- and carbon-coated grids and negatively stained with 4% uranyl acetate for TEM observation. MamK polymerized into well-developed, filamentous, straight bundles (Fig. 1.8). The bundles were comprised of fine filaments, smaller than 20 nm wide. Scanning TEM images of the filaments revealed that each filament was made up of two small helical filaments (Fig. 1.8). The mean crossover distance of the two helical filaments was 23 nm, and the helical filament was approximately 6 nm wide. Additionally, the helical filament was thus presumed to be a protofilament constructed with a single strand of MamK molecules. In vitro polymerized MamK filaments were observed using MamK protein from different species, *M. gryphiswaldense* MSR-1 (Sonkaria et al. 2012), *M. magneticum* AMB-1 (Ozyamak et al. 2013), and *Ca. Magnetobacterium casensis* (Deng et al. 2016). Recently, a high-resolution structure (~6.5 Å) of the MamK filament was determined using cryo-electron microscopy. Accordingly, the MamK filament was of double-stranded and non-staggered architecture, with monomeric MamK protein contacts apparent (Bergeron et al. 2016). The longitudinal contacts along each MamK strand most closely resembled those of eukaryotic actin; however, the cross-strand interface is novel among actin-like proteins and is responsible for the non-staggered architecture (Bergeron et al. 2016).

1.2.7 Recent Advances in Understanding the MamK Cytoskeleton

Scheffel et al. reported that *M. gryphiswaldense* MSR-1 cells in which the acidic magnetosome protein MamJ was deleted did not assemble linear chains of magnetosomes. Rather, magnetosomes aggregated into three-dimensional clusters (Scheffel et al. 2006). Furthermore, using cryo-electron tomography, they demonstrated that in Δ*mamJ* cells, magnetosome vesicles were detached from cytoskeletal filaments. These observations led to the development of a model in which MamJ connects magnetosome vesicles to the MamK filament, which serves as a scaffold for stabilizing magnetosome chains (Scheffel et al. 2006). The results of a bacterial two-hybrid study indicated that MamJ interacts with MamK in vivo, and the C- and N-terminal sequence regions of MamJ mediated these interactions (Scheffel and Schüler 2007).

Draper et al. demonstrated that *M. magneticum* AMB-1 cells deficient in both MamJ and LimJ, a paralog of MamJ, formed a chain of magnetosomes in contrast to the aggregated magnetosomes observed in *M. gryphiswaldense* MSR-1 (Draper et al. 2011). Cryo-electron tomography observations of the Δ*mamJ* Δ*limJ* strain revealed chains with 100- to 250-nm gaps, which were devoid of magnetosomes (Draper et al. 2011). Interestingly, the distribution of magnetosome-associated MamK filaments was drastically altered in the Δ*mamJ* Δ*limJ* strain. To note, the filaments were observed as concentrated bundles within the gaps of the magnetosome chains. Furthermore, using the fluorescence recovery after photobleaching (FRAP) assay, they demonstrated that MamK forms dynamic filaments in *M. magneticum* AMB-1 (Draper et al. 2011). MamK mutants of essential residues for ATP hydrolysis activity ($\text{MamK}^{\text{D161A}}$ and $\text{Mam}^{\text{K143A}}$) did not display such dynamics, thereby demonstrating that MamK ATPase activity is required for own dynamics. Strikingly, in the Δ*mamJ* Δ*limJ* strain, MamK-GFP was static as assessed by FRAP, suggesting that MamJ and LimJ promote MamK filament dynamics. Taken together, these results indicate that MamK dynamics require MamJ and LimJ for the assembly and maintenance of the magnetosome chain.

Philippe and Wu, using a bimolecular fluorescence complementation assay, demonstrated that Amb0994, the MAI-encoded methyl-accepting chemotaxis protein, interacts with MamK at the cell poles where methyl-accepting chemotaxis proteins usually cluster (Philippe and Wu 2010). They also demonstrated that Amb0994 impacts effective magnetic sensing of *M. magneticum* AMB-1 (Zhu et al. 2014). They proposed that this interaction plays a key role in magnetotaxis. Moreover, two-hybrid genomic screenings suggested that FliM, a flagella motor switch protein; Amb0854, another methyl-accepting chemotaxis protein; and Amb3568 and GGDEF domain-containing protein interact with MamK (Pan et al. 2012). These findings may lead to a novel hypothesis for magnetotaxis.

1.3 Conclusions

In the past, bacterial cells were thought of as "bags of enzymes." However, bacteria have been eloquently shown to possess highly ordered, compartmented organelles purposed for specific functions such as photosynthesis, nitrogen metabolism, and magnetic sensing. However, little is known about the molecular mechanisms governing the biogenesis and function of these bacterial organelles. Interestingly, the magnetosome possesses many features of a eukaryotic organelle. Particularly, the magnetosome is a membrane-enclosed subcellular compartment equipped with specialized proteins and cytoskeletal filaments to facilitate magnetic sensing. Additionally, magnetosomes are passed on to daughter bacterial cells. Thus, the magnetosome is an outstanding model to elucidate the molecular landscape of bacterial organelle organization. Further studies about the specific roles of magnetosome-associated proteins are warranted to shed light on magnetosome biogenesis and function.

Acknowledgments This work was supported by MEXT KAKENHI (24117007 and 23111508), JSPS KAKENHI (17H03791, 16 K07661, 25850051, and 22780063), the Human Frontier Science Program (RGP0035/2004-C104), and the Institute for Fermentation, Osaka (IFO).

References

Abreu F, Martins JL, Silveira TS, Keim CN, de Barros HG, Filho FJ, Lins U (2007) 'Candidatus Magnetoglobus multicellularis', a multicellular, magnetotactic prokaryote from a hypersaline environment. Int J Syst Evol Microbiol 57(Pt 6):1318–1322. https://doi.org/10.1099/ijs.0.64857-0

Bazylinski DA, Frankel RB (2004) Magnetosome formation in prokaryotes. Nat Rev Microbiol 2(3):217–230. https://doi.org/10.1038/nrmicro842

Bazylinski DA, Frankel RB, Jannasch HW (1988) Anaerobic magnetite production by a marine, magnetotactic bacterium. Nature 334:518–519

Bellini S (2009a) Further studies on "magnetosensitive bacteria". Chin J Oceanol Limnol 27(1):6–12. https://doi.org/10.1007/s00343-009-0006-2

Bellini S (2009b) On a unique behavior of freshwater bacteria. Chin J Oceanol Limnol 27(1):3–5. https://doi.org/10.1007/s00343-009-0003-5

Bergeron JR, Hutto R, Ozyamak E, Hom N, Hansen J, Draper O, Byrne ME, Keyhani S, Komeili A, Kollman JM (2016) Structure of the magnetosome-associated actin-like MamK filament at subnanometer resolution. Protein Sci. https://doi.org/10.1002/pro.2979

Blakemore R (1975) Magnetotactic bacteria. Science 190(4212):377–379

Blakemore RP (1982) Magnetotactic bacteria. Annu Rev Microbiol 36:217–238. https://doi.org/10.1146/annurev.mi.36.100182.001245

Blakemore RP, Maratea D, Wolfe RS (1979) Isolation and pure culture of a freshwater magnetic spirillum in chemically defined medium. J Bacteriol 140(2):720–729

Bork P, Sander C, Valencia A (1992) An ATPase domain common to prokaryotic cell-cycle proteins, sugar kinases, actin, and Hsp70 heat-shock proteins. Proc Natl Acad Sci U S A 89(16):7290–7294. https://doi.org/10.1073/pnas.89.16.7290

Deng A, Lin W, Shi N, Wu J, Sun Z, Sun Q, Bai H, Pan Y, Wen T (2016) In vitro assembly of the bacterial actin protein MamK from '*Candidatus* Magnetobacterium casensis' in the phylum Nitrospirae. Protein Cell 7(4):267–280. https://doi.org/10.1007/s13238-016-0253-x

Derman AI, Becker EC, Truong BD, Fujioka A, Tucey TM, Erb ML, Patterson PC, Pogliano J (2009) Phylogenetic analysis identifies many uncharacterized actin-like proteins (Alps) in bacteria: regulated polymerization, dynamic instability and treadmilling in Alp7A. Mol Microbiol 73(4):534–552. https://doi.org/10.1111/j.1365-2958.2009.06771.x

Draper O, Byrne ME, Li Z, Keyhani S, Barrozo JC, Jensen G, Komeili A (2011) MamK, a bacterial actin, forms dynamic filaments in vivo that are regulated by the acidic proteins MamJ and LimJ. Mol Microbiol 82(2):342–354. https://doi.org/10.1111/j.1365-2958.2011.07815.x

Dufrêne YF (2008) Towards nanomicrobiology using atomic force microscopy. Nat Rev Microbiol 6(9):674–680. https://doi.org/10.1038/nrmicro1948

Engel A, Müller DJ (2000) Observing single biomolecules at work with the atomic force microscope. Nat Struct Biol 7(9):715–718. https://doi.org/10.1038/78929

Faivre D, Schüler D (2008) Magnetotactic bacteria and magnetosomes. Chem Rev 108(11):4875–4898. https://doi.org/10.1021/cr078258w

Flies CB, Peplies J, Schüler D (2005) Combined approach for characterization of uncultivated magnetotactic bacteria from various aquatic environments. Appl Environ Microbiol 71(5):2723–2731. https://doi.org/10.1128/Aem.71.5.2723-2731.2005

Frankel RB (2009) The discovery of magnetotactic/magnetosensitive bacteria. Chin J Oceanol Limnol 27(1):1–2. https://doi.org/10.1007/s00343-009-0001-7

Frankel RB, Bazylinski DA, Johnson MS, Taylor BL (1997) Magneto-aerotaxis in marine coccoid bacteria. Biophys J 73(2):994–1000

Fukuda Y, Okamura Y, Takeyama H, Matsunaga T (2006) Dynamic analysis of a genomic island in *Magnetospirillum* sp. strain AMB-1 reveals how magnetosome synthesis developed. FEBS Lett 580(3):801–812. https://doi.org/10.1016/j.febstet.2006.01.003

Gorlenko VM, Dzyuba MV, Maleeva AN, Panteleeva AN, Kolganova TV, Kuznetsov BB (2011) *Magnetospirillum aberrantis* sp. nov., a new freshwater bacterium with magnetic inclusions. Microbiology 80(5):692–702. https://doi.org/10.1134/S0026261711050055

Jogler C, Lin W, Meyerdierks A, Kube M, Katzmann E, Flies C, Pan YX, Amann R, Reinhardt R, Schüler D (2009) Toward cloning of the magnetotactic metagenome: identification of magnetosome island gene clusters in uncultivated magnetotactic bacteria from different aquatic sediments. Appl Environ Microbiol 75(12):3972–3979. https://doi.org/10.1128/Aem.02701-08

Jones LJF, Carballido-López R, Errington J (2001) Control of cell shape in bacteria: helical, actin-like filaments in *Bacillus subtilis*. Cell 104(6):913–922. https://doi.org/10.1016/S0092-8674(01)00287-2

Kolinko S, Jogler C, Katzmann E, Wanner G, Peplies J, Schüler D (2012) Single-cell analysis reveals a novel uncultivated magnetotactic bacterium within the candidate division OP3. Environ Microbiol 14(7):1709–1721. https://doi.org/10.1111/j.1462-2920.2011.02609.x

Kolinko S, Wanner G, Katzmann E, Kiemer F, Fuchs BM, Schüler D (2013) Clone libraries and single cell genome amplification reveal extended diversity of uncultivated magnetotactic bacteria from marine and freshwater environments. Environ Microbiol 15(5):1290–1301. https://doi.org/10.1111/1462-2920.12004

Komeili A (2012) Molecular mechanisms of compartmentalization and biomineralization in magnetotactic bacteria. FEMS Microbiol Rev 36(1):232–255. https://doi.org/10.1111/j.1574-6976.2011.00315.x

Komeili A, Li Z, Newman DK, Jensen GJ (2006) Magnetosomes are cell membrane invaginations organized by the actin-like protein MamK. Science 311(5758):242–245. https://doi.org/10.1126/science.1123231

Kruse T, Møller-Jensen J, Løbner-Olesen A, Gerdes K (2003) Dysfunctional MreB inhibits chromosome segregation in *Escherichia coli*. EMBO J 22(19):5283–5292. https://doi.org/10.1093/emboj/cdg504

Lefèvre CT, Bazylinski DA (2013) Ecology, diversity, and evolution of magnetotactic bacteria. Microbiol Mol Biol Rev: MMBR 77(3):497–526. https://doi.org/10.1128/MMBR.00021-13
Lefèvre CT, Wu LF (2013) Evolution of the bacterial organelle responsible for magnetotaxis. Trends Microbiol 21(10):534–543. https://doi.org/10.1016/j.tim.2013.07.005
Lefèvre CT, Bernadac A, Yu-Zhang K, Pradel N, Wu LF (2009) Isolation and characterization of a magnetotactic bacterial culture from the Mediterranean Sea. Environ Microbiol 11(7):1646–1657. https://doi.org/10.1111/j.1462-2920.2009.01887.x
Lefèvre CT, Abreu F, Schmidt ML, Lins U, Frankel RB, Hedlund BP, Bazylinski DA (2010) Moderately thermophilic magnetotactic bacteria from hot springs in Nevada. Appl Environ Microbiol 76(11):3740–3743. https://doi.org/10.1128/Aem.03018-09
Lefèvre CT, Frankel RB, Abreu F, Lins U, Bazylinski DA (2011a) Culture-independent characterization of a novel, uncultivated magnetotactic member of the *Nitrospirae* phylum. Environ Microbiol 13(2):538–549. https://doi.org/10.1111/j.1462-2920.2010.02361.x
Lefèvre CT, Frankel RB, Pósfai M, Prozorov T, Bazylinski DA (2011b) Isolation of obligately alkaliphilic magnetotactic bacteria from extremely alkaline environments. Environ Microbiol 13(8):2342–2350. https://doi.org/10.1111/j.1462-2920.2011.02505.x
Lefèvre CT, Menguy N, Abreu F, Lins U, Pósfai M, Prozorov T, Pignol D, Frankel RB, Bazylinski DA (2011c) A cultured greigite-producing magnetotactic bacterium in a novel group of sulfate-reducing bacteria. Science 334(6063):1720–1723. https://doi.org/10.1126/science.1212596
Lefèvre CT, Viloria N, Schmidt ML, Posfai M, Frankel RB, Bazylinski DA (2012) Novel magnetite-producing magnetotactic bacteria belonging to the *Gammaproteobacteria*. ISME J 6(2):440–450. https://doi.org/10.1038/ismej.2011.97
Lefèvre CT, Trubitsyn D, Abreu F, Kolinko S, de Almeida LGP, de Vasconcelos ATR, Lins U, Schüler D, Ginet N, Pignol D, Bazylinski DA (2013) Monophyletic origin of magnetotaxis and the first magnetosomes. Environ Microbiol 15(8):2267–2274. https://doi.org/10.1111/1462-2920.12097
Li WB, Yu LJ, Zhou PP, Min Z (2007) Isolation of magnetotactic bacterium WM-1 from freshwater sediment and phylogenetic characterization. Arch Microbiol 188(1):97–102. https://doi.org/10.1007/s00203-007-0231-z
Lin W, Li JH, Pan YX (2012) Newly isolated but uncultivated magnetotactic bacterium of the phylum *Nitrospirae* from Beijing, China. Appl Environ Microbiol 78(3):668–675. https://doi.org/10.1128/Aem.06764-11
Lin W, Bazylinski DA, Xiao T, Wu LF, Pan Y (2013) Life with compass: diversity and biogeography of magnetotactic bacteria. Environ Microbiol 16(9):2646–2658. https://doi.org/10.1111/1462-2920.12313
Lin W, Deng AH, Wang Z, Li Y, Wen TY, Wu LF, Wu M, Pan YX (2014) Genomic insights into the uncultured genus '*Candidatus* Magnetobacterium' in the phylum *Nitrospirae*. ISME J 8(12):2463–2477. https://doi.org/10.1038/ismej.2014.94
Lohße A, Ullrich S, Katzmann E, Borg S, Wanner G, Richter M, Voigt B, Schweder T, Schüler D (2011) Functional analysis of the magnetosome island in *Magnetospirillum gryphiswaldense*: the *mamAB* operon is sufficient for magnetite biomineralization. PLoS One 6(10):e25561. https://doi.org/10.1371/journal.pone.0025561
Matsunaga T, Sakaguchi T, Tadokoro F (1991) Magnetite formation by a magnetic bacterium capable of growing aerobically. Appl Microbiol Biotechnol 35(5):651–655
Matsunaga T, Okamura Y, Fukuda Y, Wahyudi AT, Murase Y, Takeyama H (2005) Complete genome sequence of the facultative anaerobic magnetotactic bacterium *Magnetospirillum* sp. strain AMB-1. DNA Res 12(3):157–166. https://doi.org/10.1093/dnares/dsi002
Meldrum FC, Mann S, Heywood BR, Frankel RB, Bazylinski DA (1993a) Electron microscopy study of magnetosomes in two cultured vibrioid magnetotactic bacteria. Proc R Soc Lond B Biol Sci 251(1332):237–242
Meldrum FC, Mann S, Heywood BR, Frankel RB, Bazylinski DA (1993b) Electron-microscopy study of magnetosomes in a cultured coccoid magnetotactic bacterium. Proc R Soc B-Biol Sci 251(1332):231–236. https://doi.org/10.1098/rspb.1993.0034

Murat D, Quinlan A, Vali H, Komeili A (2010) Comprehensive genetic dissection of the magnetosome gene island reveals the step-wise assembly of a prokaryotic organelle. Proc Natl Acad Sci U S A 107(12):5593–5598. https://doi.org/10.1073/pnas.0914439107

Nguyen HV, Suzuki E, Oestreicher Z, Minamide H, Endoh H, Fukumori Y, Taoka A (2016) A protein-protein interaction in magnetosomes: TPR protein MamA interacts with an Mms6 protein. Biochem Biophys Rep 7:39–44. https://doi.org/10.1016/j.bbrep.2016.05.010

Okuda Y, Fukumori Y (2001) Expression and characterization of a magnetosome-associated protein, TPR-containing MAM22, in *Escherichia coli*. FEBS Lett 491(3):169–173. https://doi.org/10.1016/S0014-5793(01)02178-0

Okuda Y, Denda K, Fukumori Y (1996) Cloning and sequencing of a gene encoding a new member of the tetratricopeptide protein family from magnetosomes of *Magnetospirillum magnetotacticum*. Gene 171(1):99–102. https://doi.org/10.1016/0378-1119(95)00008-9

Ozyamak E, Kollman J, Agard DA, Komeili A (2013) The bacterial actin MamK: in vitro assembly behavior and filament architecture. J Biol Chem 288(6):4265–4277. https://doi.org/10.1074/jbc.M112.417030

Pan WD, Xie CL, Lv J (2012) Screening for the interacting partners of the proteins MamK & MamJ by two-hybrid genomic DNA library of *Magnetospirillum magneticum* AMB-1. Curr Microbiol 64(6):515–523. https://doi.org/10.1007/s00284-012-0099-2

Petersen N, Weiss DG, Vali H (1989) Magnetic bacteria in lake sediments. Geomagnetism and palaeomagnetism, The NATO ASI Series, vol 261. Springer, Berlin, pp 231–241

Philippe N, Wu LF (2010) An MCP-like protein interacts with the MamK cytoskeleton and is involved in magnetotaxis in *Magnetospirillum magneticum* AMB-1. J Mol Biol 400(3):309–322. https://doi.org/10.1016/j.jmb.2010.05.011

Revathy T, Jacob JJ, Jayasri MA, Suthindhiran K (2016) Isolation and characterization of *Magnetospirillum* from saline lagoon. World J Microbiol Biotechnol 32(7):109. https://doi.org/10.1007/s11274-016-2075-7

Richter M, Kube M, Bazylinski DA, Lombardot T, Glöckner FO, Reinhardt R, Schüler D (2007) Comparative genome analysis of four magnetotactic bacteria reveals a complex set of group-specific genes implicated in magnetosome biomineralization and function. J Bacteriol 189(13):4899–4910. https://doi.org/10.1128/Jb.00119-07

Sakaguchi T, Burgess JG, Matsunaga T (1993) Magnetite formation by a sulfate-reducing bacterium. Nature 365(6441):47–49. https://doi.org/10.1038/365047a0

Scheffel A, Schüler D (2007) The acidic repetitive domain of the *Magnetospirillum gryphiswaldense* MamJ protein displays hypervariability but is not required for magnetosome chain assembly. J Bacteriol 189(17):6437–6446. https://doi.org/10.1128/Jb.00421-07

Scheffel A, Gruska M, Faivre D, Linaroudis A, Plitzko JM, Schüler D (2006) An acidic protein aligns magnetosomes along a filamentous structure in magnetotactic bacteria. Nature 440(7080):110–114. https://doi.org/10.1038/nature04382

Schleifer KH, Schüler D, Spring S, Weizenegger M, Amann R, Ludwig W, Kohler M (1991) The genus *Magnetospirillum* gen. nov. description of *Magnetospirillum gryphiswaldense* sp. nov. and transfer of *Aquaspirillum magnetotacticum* to *Magnetospirillum magnetotacticum* comb. nov. Syst Appl Microbiol 14(4):379–385

Schübbe S, Kube M, Scheffel A, Wawer C, Heyen U, Meyerdierks A, Madkour MH, Mayer F, Reinhardt R, Schüler D (2003) Characterization of a spontaneous nonmagnetic mutant of *Magnetospirillum gryphiswaldense* reveals a large deletion comprising a putative magnetosome island. J Bacteriol 185(19):5779–5790. https://doi.org/10.1128/Jb.185.19.5779-5790.2003

Schüler D (2008) Genetics and cell biology of magnetosome formation in magnetotactic bacteria. FEMS Microbiol Rev 32(4):654–672. https://doi.org/10.1111/j.1574-6976.2008.00116.x

Shaevitz JW, Gitai Z (2010) The structure and function of bacterial actin homologs. Cold Spring Harb Perspect Biol 2(9):a000364. https://doi.org/10.1101/cshperspect.a000364

Sonkaria S, Fuentes G, Verma C, Narang R, Khare V, Fischer A, Faivre D (2012) Insight into the assembly properties and functional organisation of the magnetotactic bacterial actin-like homolog, MamK. PLoS One 7(5):e34189. https://doi.org/10.1371/journal.pone.0034189

Taoka A, Asada R, Sasaki H, Anzawa K, Wu L, Fukumori Y (2006) Spatial localizations of Mam22 and Mam12 in the magnetosomes of *Magnetospirillum magnetotacticum*. J Bacteriol 188(11):3805–3812. https://doi.org/10.1128/JB.00020-06

Taoka A, Asada R, Wu L-F, Fukumori Y (2007) Polymerization of the actin-like protein MamK, which is associated with magnetosomes. J Bacteriol 189(23):8737–8740. https://doi.org/10.1128/JB.00899-07

Taoka A, Kondo J, Oestreicher Z, Fukumori Y (2014) Characterization of uncultured giant rod-shaped magnetotactic *Gammaproteobacteria* from a freshwater pond in Kanazawa, Japan. Microbiol-SGM 160:2226–2234. https://doi.org/10.1099/mic.0.078717-0

Thrash JC, Ahmadi S, Torok T, Coates JD (2010) *Magnetospirillum bellicus* sp. nov., a novel dissimilatory perchlorate-reducing Alphaproteobacterium isolated from a bioelectrical reactor. Appl Environ Microbiol 76(14):4730–4737. https://doi.org/10.1128/Aem.00015-10

Ullrich S, Kube M, Schübbe S, Reinhardt R, Schüler D (2005) A hypervariable 130-kilobase genomic region of *Magnetospirillum gryphiswaldense* comprises a magnetosome island which undergoes frequent rearrangements during stationary growth. J Bacteriol 187(21):7176–7184. https://doi.org/10.1128/Jb.187.21.7176-7184.2005

van den Ent F, Amos LA, Löwe J (2001) Prokaryotic origin of the actin cytoskeleton. Nature 413(6851):39–44. https://doi.org/10.1038/35092500

van den Ent F, Møller-Jensen J, Amos LA, Gerdes K, Löwe J (2002) F-actin-like filaments formed by plasmid segregation protein ParM. EMBO J 21(24):6935–6943. https://doi.org/10.1093/emboj/cdf672

Wang YZ, Lin W, Li JH, Zhang TW, Li Y, Tian JS, Gu LX, Heyden YV, Pan YX (2015) Characterizing and optimizing magnetosome production of *Magnetospirillum* sp XM-1 isolated from Xi'an City Moat, China. FEMS Microbiol Lett 362(21):fnv167. https://doi.org/10.1093/femsle/fnv167

Wenter R, Wanner G, Schüler D, Overmann J (2009) Ultrastructure, tactic behaviour and potential for sulfate reduction of a novel multicellular magnetotactic prokaryote from North Sea sediments. Environ Microbiol 11(6):1493–1505. https://doi.org/10.1111/j.1462-2920.2009.01877.x

Williams TJ, Lefèvre CT, Zhao WD, Beveridge TJ, Bazylinski DA (2012) *Magnetospira thiophila* gen. nov., sp. nov., a marine magnetotactic bacterium that represents a novel lineage within the *Rhodospirillaceae* (*Alphaproteobacteria*). Int J Syst Evol Microbiol 62:2443–2450. https://doi.org/10.1099/ijs.0.037697-0

Yamamoto D, Taoka A, Uchihashi T, Sasaki H, Watanabe H, Ando T, Fukumori Y (2010) Visualization and structural analysis of the bacterial magnetic organelle magnetosome using atomic force microscopy. Proc Natl Acad Sci U S A 107(20):9382–9387. https://doi.org/10.1073/pnas.1001870107

Zeytuni N, Zarivach R (2012) Structural and functional discussion of the tetra-trico-peptide repeat, a protein interaction module. Structure 20(3):397–405. https://doi.org/10.1016/j.str.2012.01.006

Zeytuni N, Ozyamak E, Ben-Harush K, Davidov G, Levin M, Gat Y, Moyal T, Brik A, Komeili A, Zarivach R (2011) Self-recognition mechanism of MamA, a magnetosome-associated TPR-containing protein, promotes complex assembly. Proc Natl Acad Sci U S A 108(33):E480–E487. https://doi.org/10.1073/pnas.1103367108

Zeytuni N, Baran D, Davidov G, Zarivach R (2012) Inter-phylum structural conservation of the magnetosome-associated TPR-containing protein, MamA. J Struct Biol 180(3):479–487. https://doi.org/10.1016/j.jsb.2012.08.001

Zeytuni N, Cronin S, Lefèvre CT, Arnoux P, Baran D, Shtein Z, Davidov G, Zarivach R (2015) MamA as a model protein for structure-based insight into the evolutionary origins of magnetotactic bacteria. PLoS One 10(6):e0130394. https://doi.org/10.1371/journal.pone.0130394

Zhang WY, Zhou K, Pan HM, Yue HD, Jiang M, Xiao T, Wu LF (2012) Two genera of magnetococci with bean-like morphology from intertidal sediments of the yellow sea, China. Appl Environ Microbiol 78(16):5606–5611. https://doi.org/10.1128/Aem.00081-12

Zhou K, Zhang WY, Yu-Zhang K, Pan HM, Zhang SD, Zhang WJ, Yue HD, Li Y, Xiao T, Wu LF (2012) A novel genus of multicellular magnetotactic prokaryotes from the Yellow Sea. Environ Microbiol 14(2):405–413. https://doi.org/10.1111/j.1462-2920.2011.02590.x

Zhu KL, Pan HM, Li JH, Yu-Zhang K, Zhang SD, Zhang WY, Zhou K, Yue HD, Pan YX, Xiao TA, Wu LF (2010) Isolation and characterization of a marine magnetotactic spirillum axenic culture QH-2 from an intertidal zone of the China Sea. Res Microbiol 161(4):276–283. https://doi.org/10.1016/j.resmic.2010.02.003

Zhu X, Ge X, Li N, Wu LF, Luo C, Ouyang Q, Tu Y, Chen G (2014) Angle sensing in magnetotaxis of *Magnetospirillum magneticum* AMB-1. Integr Biol 6(7):706–713. https://doi.org/10.1039/c3ib40259b

Chapter 2
Molecular Mechanism of Magnetic Crystal Formation in Magnetotactic Bacteria

Atsushi Arakaki, Masayoshi Tanaka, and Tadashi Matsunaga

Abstract Magnetotactic bacteria are a group of microorganism producing nano-sized magnetic particles. The bacterial cells accumulate a large amount of iron ion from aquatic environment and synthesize single-crystal magnetic nano-particles under ambient conditions. The size, shape, and composition of the magnetic nano-particles are precisely regulated in individual bacterial cell types. Thus, the understanding of molecular mechanism should provide ideas to design and create magnetic nano-materials with environmentally friendly synthetic routes. This chapter describes the molecular mechanism of magnetic nano-particle formation that has been clarified based on comprehensive molecular analyses of magnetotactic bacteria. Identified proteins from the basic studies are shown to be available for the development of novel magnetic nano-materials. The strategy and fundamental technologies that are useful for the understanding of biomineralization mechanisms are also introduced.

2.1 Introduction

Magnetotactic bacteria produce finely tuned nano-sized single-crystal magnetic particles consisting of Fe_3O_4 (magnetite) or Fe_3S_4 (greigite) (Blakemore 1975; Bazylinski et al. 1995). The nano-particles are approximately 40–100 nm in diameter, which is the size range of magnetically stable single-domain size. In the bacterial cell, they are naturally covered with lipid bilayer membrane (Gorby et al. 1988).

A. Arakaki · T. Matsunaga (✉)
Division of Biotechnology and Lifescience, Institute of Engineering, Tokyo University of Agriculture and Technology, Tokyo, Japan
e-mail: tmatsuna@cc.tuat.ac.jp

M. Tanaka
Division of Biotechnology and Lifescience, Institute of Engineering, Tokyo University of Agriculture and Technology, Tokyo, Japan

Department of Chemical Science and Engineering, School of Materials and Chemical Technology, Tokyo Institute of Technology, Tokyo, Japan

T. Matsunaga et al. (eds.), *Biological Magnetic Materials and Applications*, https://doi.org/10.1007/978-981-10-8069-2_2

Because of the presence of lipid membrane, the magnetic nano-particles show superior dispersibility in solution when they are extracted from the bacterial cells. In addition, to use the surface membrane as a basement, functional molecules can be attached to the magnetic nano-particle surface (Matsunaga and Kamiya 1987). Thus, surface-modified magnetic nano-particles are shown to be useful for various biotechnological applications (Arakaki et al. 2008), although the details will be described in the other chapters of this book (see Sects. 2.1 and 2.2).

By the development of genetic techniques and clarification of genome information, our knowledge on the formation mechanism of bacterial magnetic nano-particles has significantly been improved over the past decade. The number of genes and proteins involved in the nano-particle formation has been identified by comprehensive molecular analyses including comparative genome and proteome analyses (Grünberg et al. 2004; Matsunaga et al. 2005, 2009; Tanaka et al. 2006; Nakazawa and Arakaki 2009; Schübbe et al. 2009; Lefèvre et al. 2013). Their functions are then characterized based on phenotypic studies of the gene deletion mutants and biochemical studies on the proteins (see Sect. 2.3). The molecular studies revealed that the mechanism is unique in this organism and quite complicated system as consisting of several formation processes with number of biological molecules (see Sect. 2.3). The biological system enables magnetotactic bacteria to synthesize single-crystal magnetic nano-particles under the ambient room temperature and atmospheric condition.

This chapter describes the synthetic principle of magnetic nano-particles in magnetotactic bacteria, as an example of material synthesis in nature. Because the membrane-encapsulated magnetic nano-particles formed in the bacteria specifically referred to as magnetosomes, this term is used throughout this chapter. The mechanism involving multisteps with the aid of biological molecules will provide ideas to design new synthetic route of iron oxide nano-materials and develop novel magnetic nano-materials to mimic or utilize the biological materials.

2.2 Identification of Key Biological Molecular Components for Magnetic Nano-particle Formation

Comprehensive analyses such as genome sequencing (genome analysis), protein expression analysis (proteomics), and gene expression analysis (trasncriptomics) are currently recognized as powerful tools for the understanding of biological system. Genomics can list up all the molecular components of the analyzed organism. Proteomics can identify molecules that are really functional in the cell. It also provides localization information of the molecules within the cell. Transcriptomics can provide magnitude of expression levels of the molecules and also visualize their expression networks. Combinatorial use of these analyses is informative toward the identification of key molecules, and thus they are generally used as a first examination for overall understanding of organisms.

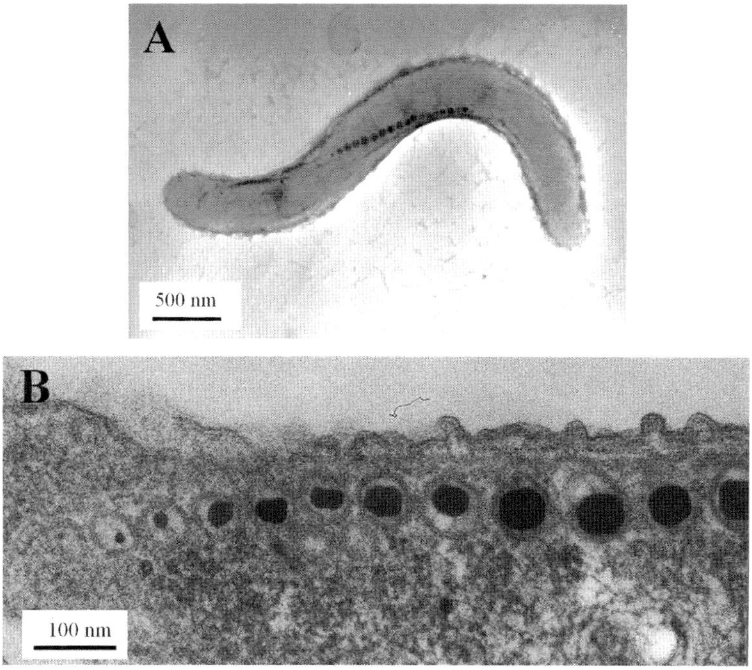

Fig. 2.1 TEM picture (**a**) and thin section (**b**) of *Magnetospirillum magneticum* AMB-1

As described in Sect. 1.1, overall mechanism of magnetosome formation was speculated from microscopic observations, microbiological characterizations, and biochemical analyses of magnetotactic bacteria. Since magnetotactic bacteria are commonly observed in microaerobic or anaerobic aquatic environments, it is considered that the organisms uptake iron as a form of Fe (II) or Fe (III) ions from the water and crystallize them to form magnetic nano-particle in magnetosome (Nakamura et al. 1995; Suzuki et al. 2006). Thin-section transmission electron microscope (TEM) observation revealed the presence of membranous structures on the surface of magnetic nano-particles (Gorby et al. 1988). The observed structure called as magnetosome membrane mainly consists of phospholipids and proteins, suggesting a presence of bacterial organelle specialized for magnetic nano-particle formation (Nakamura et al. 1991). In order to identify key molecules for the magnetosome formation, the above-described comprehensive analyses were simultaneously conducted. The first complete genome analysis of magnetotactic bacteria was achieved in a model organism, *Magnetospirillum magneticum* strain AMB-1 (Fig. 2.1), in 2005 (Matsunaga et al. 2005). The sequence has been conducted by the traditional DNA sequencing method to use capillary sequencers, and the genome size was precisely determined as 4,967,148 bp (Table 2.1). The genome data revealed the presence of 4559 predicted genes. The gene numbers are relatively smaller compared with the genome size, suggesting presence of noncoding regions in the genome. In fact, the genome includes numbers of vestiges of past exogenous

Table 2.1 Comparison of the genome characteristics of magnetotactic bacteria

Genus, species, and strains	Magnetic particle types	DNA types	Length (bp)	Predicted gene numbers	References
Magnetospirillum magneticum AMB-1	Cubo-octahedral magnetite	Chromosome	4,967,148	4559	Matsunaga et al. (2005)
Magnetospirillum magnetotacticum MS-1	Cubo-octahedral magnetite	Chromosome	4,523,935	4136	Smalley et al. (2015)
Magnetospirillum gryphiswaldense MSR-1	Cubo-octahedral magnetite	Chromosome	4,365,796	4261	Wang et al. (2014)
		Plasmid	35,803	38	
Magnetococcus marinus MC-1	Pseudohexagonal prismatic magnetite	Chromosome	4,719,581	3815	Schübbe et al. (2009)
Magnetospira sp. QH-2	Magnetite	Chromosome	4,020,900	3794	Ji et al. (2014)
		Plasmid	31,063	37	
Magnetovibrio blakemorei MV-1	Truncated hexa-octahedral magnetite	Chromosome	Approx. 3.7 M	–	Bazylinski et al. (2013)
Desulfovibrio magneticus RS-1	Bullet-shaped magnetite	Chromosome	5,248,049	4629	Nakazawa et al. (2009)
		Plasmid 1	58,704	65	
		Plasmid 2	8867	10	
Candidatus Magnetobacterium casensis	Bullet-shaped magnetite	Chromosome	3,415,676	3140	Lin et al. (2014)
Candidatus Magnetoglobus multicellularis	Magnetite and greigite	Chromosome	12,479,246	9987	Abreu et al. (2014)
Magnetospirillum moscoviense BB-1	Magnetite	Chromosome	4,164,497	4279	Koziaeva et al. (2016)
Magnetospirillum marisnigri SP-1	Magnetite	Chromosome	4,619,819	4130	Koziaeva et al. (2016)

gene transfers, such as insertion sequence (IS) elements, integrases, and phage-coding genes (Matsunaga et al. 2005). Similar features were also found in other magnetotactic bacterial strains (Nakazawa and Arakaki 2009; Ji et al. 2014). They are considered to be involved in the development of magnetosome formation, as well as the evolution of this microorganism. After the genome sequencing of strain AMB-1, genomes of other magnetotactic bacterial strains have also become available (Table 2.1). Next-generation sequencers facilitated the analyses. New strategies enabling to analyze genomes from uncultured magnetotactic bacteria were also developed by using magnetic separation of bacteria and whole genome amplification (Arakaki et al. 2010; Jogler et al. 2011; Kolinko et al. 2012). Comparative genome analysis of both isolated and unisolated magnetotactic bacteria revealed the presence of common gene sets in magnetotactic bacteria. The gene region is specifically called as magnetosome island. The comparative analyses of magnetosome

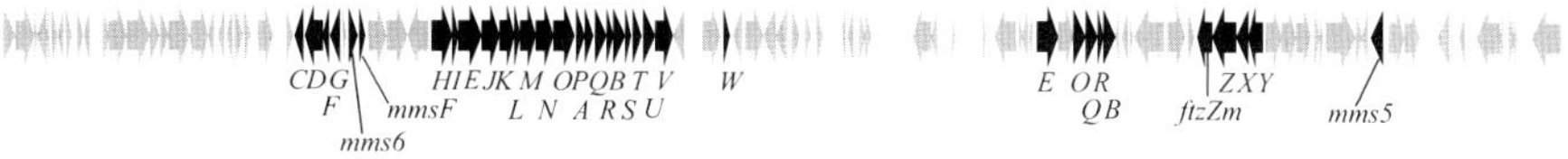

Fig. 2.2 Molecular organization of genes in magnetosome island of *M. magneticum* AMB-1. The predicted ORFs are represented by arrowheads indicating the direction of transcription. Black arrowheads indicate genes encoding magnetosome proteins. *mamA–mamZ* genes are abbreviated as *A–Z*

islands revealed the distinct variations in gene order and sequence similarity of the genes in these microorganisms (Richter et al. 2007; Nakazawa and Arakaki 2009; Lefèvre and Wu 2013). As an example, the magnetosome island of strain AMB-1 is shown in Fig. 2.2. Spontaneous deletion of this region from the AMB-1 genome resulted deficiency of magnetosome formation ability in the mutant strain (Fukuda et al. 2006). The findings suggested the presence of core genetic components for magnetosome formation; these genes may have been acquired into the magnetotactic bacterial genomes by multiple gene transfer events during evolution of this organism.

Proteome and transcriptome analyses generally require gene sequences. Thus, the abovementioned genome information was utilized for these analyses for magnetotactic bacteria. Proteome analyses of magnetosome membrane proteins have been conducted in several species of magnetotactic bacteria from which numerous novel proteins with potentially crucial roles in magnetosome biomineralization have been discovered. In the analyses of magnetosome membrane proteins from *M. magneticum* AMB-1 and *M. gryphiswaldense* MSR-1, approximately 80 and 30 proteins were identified (Grünberg et al. 2004; Tanaka et al. 2006), respectively. A part of the proteins are listed in Table 2.2. A considerable number of identified proteins were found to be assigned in gene clusters located within the magnetosome island. Furthermore, comparative proteome analysis between the protein fractions acquired from the outer membrane, cytoplasmic membrane, magnetosome membrane, and cytoplasm-periplasmic fractions was conducted (Tanaka et al. 2006). A high degree of similarity was observed between the protein profiles of the magnetosome and cytoplasmic membranes of the AMB-1 strain. Fatty acid comparative analysis also indicated that both these fractions showed similar profiles. These results suggest that the magnetosome membrane could have been derived from the cytoplasmic membrane (Tanaka et al. 2006). Proteome analysis of *D. magneticus* RS-1 revealed a presence of magnetosome membrane proteins commonly exist in strain RS-1 and *Magnetospirillum* spp. (Matsunaga et al. 2009). Those include actin-like protein, TPR-containing protein, cation diffusion facilitator, and several oxidation-reduction proteins. On the other hand, non-common proteins were also identified from magnetosome of strain RS-1 (Matsunaga et al. 2009). The results indicate that the phylogenetically distinct magnetotactic bacteria utilize common proteins to form magnetosome, although some proteins are different.

Table 2.2 A list of proteins related to magnetosome formation

Protein	Suggested role	Protein identification		Genetic investigation[a]	Main related articles
		AMB-1	MSR-1		
	Membrane formation				
Mms16	Membrane invagination	✓	✓		Okamura et al. (2001)
MpsA	Lipid metabolism	✓			Matsunaga et al. (2000)
MamA (Mms24, Mam22)	Vesicle activation	✓	✓	HR	Komeili and Vali (2004), Yamamoto et al. (2010), Zeytuni et al. (2011, 2015), Lohße et al. (2014), and Nguyen et al. (2016)
MamI	Membrane formation			HR	Murat et al. (2010), Lohße et al. (2014), and Raschdorf et al. (2016)
MamL	Membrane formation			HR	Murat et al. 2010, Lohße et al. (2014), and Raschdorf et al. (2016)
MamQ	Membrane formation		✓	HR	Murat et al. (2010), Lohße et al. (2014), and Raschdorf et al. (2016)
MamY	Membrane constriction	✓		HR	Tanaka et al. (2010a)
	Chain formation				
MamJ	Chain formation	✓	✓	HR	Scheffel et al. (2006), Scheffel and Schüler (2007), and Pan et al. (2012)
MamK	Chain formation	✓		HR	Komeili et al. (2006), Taoka et al. (2007), Draper et al. (2011), Pan et al. (2012), and Bergeron et al. (2016)
	Iron transport and redox control				
MagA	Iron transport			TP	Matsunaga et al. (1992), Nakamura et al. (1995), and Wang et al. (2009)
MamB	Iron transport, vesicle formation	✓	✓	HR	Murat et al. (2010), Uebe et al. (2011), and Raschdorf et al. (2016)
MamH	Iron transport			HR	Murat et al. (2010), and Raschdorf et al. (2013)
MamM	Iron transport		✓	HR	Murat et al. (2010), Uebe et al. (2011), Zeytuni et al. (2014), and Raschdorf et al. (2016)
MamZ	Iron transport, redox control			HR	Raschdorf et al. (2013)

(continued)

Table 2.2 (continued)

Protein	Suggested role	Protein identification AMB-1	MSR-1	Genetic investigation[a]	Main related articles
AOR	Redox control			TP	Wahyudi et al. (2003)
MamE	Redox control, protein sorting	✓	✓	HR	Yang et al. (2010), Murat et al. (2010), Quinlan et al. 2011, and Hershey et al. (2016)
MamP	Redox control	✓		HR	Murat et al. (2010), Siponen et al. (2013), Taoka et al. (2014), Lohße et al. (2014), Jones et al. (2015), and Hershey et al. (2016)
MamT	Redox control		✓	HR	Murat et al. (2010), Siponen et al. (2012), Lohße et al. (2014), and Jones et al. (2015)
MamX	Redox control	✓		HR	Yang et al. (2013) and Raschdorf et al. (2013)
	Magnetite biomineralization				
Mms5	Crystal formation	✓		HR	Arakaki et al. (2003, 2014)
Mms6	Crystal formation	✓	✓	HR	Arakaki et al. (2003, 2014), Wang et al. (2009), Tanaka et al. (2011b), Lohße et al. (2014), and Nguyen et al. (2016)
Mms7 (MamD)	Crystal formation	✓	✓	HR, GI	Arakaki et al. (2003, 2014), Scheffel et al. (2008), and Yamagishi et al. (2016)
Mms13 (MamC)	Crystal formation	✓	✓	HR, GI	Arakaki et al. (2003, 2014), Scheffel et al. (2008), and Yoshino et al. (2010)
MamF	Crystal formation		✓	HR	Scheffel et al. (2008) and Lohße et al. (2014)
MamG	Crystal formation		✓	HR	Scheffel et al. (2008)
MamO	Nucleation		✓	HR	Yang et al. (2010), Quinlan et al. (2011), Guo et al. (2012), and Hershey et al. (2016)
FtsZm	Crystal formation			HR	Ding et al. (2010) and Müller et al. (2014)
MmsF	Crystal formation			HR	Murat and Falahati (2012), Rawlings et al. (2014), and Lohße et al. (2014)
	Function unknown				

(continued)

Table 2.2 (continued)

Protein	Suggested role	Protein identification		Genetic investigation[a]	Main related articles
		AMB-1	MSR-1		
MamN	–		✓	HR	Murat et al. (2010), Lohße et al. (2014), and Raschdorf et al. (2016)
MamU	–			HR	Murat et al. (2010) and Lohße et al. (2014)
MamV	–			HR	Murat et al. (2010)
MamR	–		✓	HR	Murat et al. (2010) and Lohße et al. (2014)
MamS	–	✓	✓	HR	Murat et al. (2010) and Lohße et al. (2014)

[a]The tools for genetic investigation were listed. The abbreviations were as follows: *TP* phenotypic analysis of mutant strain developed by transposon mutagenesis, *HR* phenotypic analysis of mutant strain developed by homologous recombination, and *GI* gene expression induction

Transcriptome analysis supported results from genome and proteome analyses. The analysis is generally conducted to use DNA microarray and real-time reverse transcription PCR (RT-PCR). Since magnetotactic bacteria uptake a larger amount of iron compared to other well-characterized bacteria, global gene expression analysis of strain AMB-1 grown under iron-rich and iron-deficient conditions was conducted (Suzuki et al. 2006). The result indicated that despite the unusual high-iron requirement of strain AMB-1, it utilizes robust but simple iron uptake systems similar to those of other gram-negative bacteria. Analysis of the genes in magnetosome island revealed that the genes were transcribed under magnetosome-forming conditions, while they were downregulated under non-forming conditions (Schübbe et al. 2006).

The comprehensive analyses successfully clarified genes and proteins for magnetic nano-particle formation in magnetotactic bacteria. A recent study revealed that non-magnetotactic photosynthetic bacterium can produce magnetosome by inserting the genes of magnetosome island into the genome (Kolinko et al. 2014). To use the molecular components identified here, a genetically modified microorganism which can produce both magnetic nano-particles and other useful substances will be constructed in near future.

2.3 Molecular Analysis of Unique Prokaryotic Organelle Biogenesis Specialized for Magnetic Nano-particle Production

Various magnetotactic bacteria have been isolated and studied in many research groups. Among these magnetotactic bacteria, *M. magneticum* AMB-1, *M. gryphiswaldense* MSR-1, *M. magnetotacticum* strain MS-1, *Magnetococcus marinus* strain MC-1, and *Desulfovibrio magneticus* RS-1 are the most characterized strains.

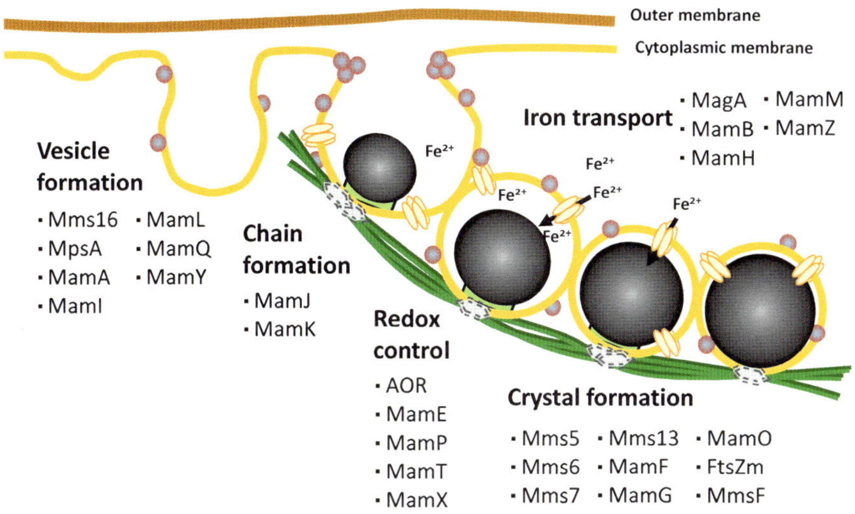

Fig. 2.3 Schematic illustration of hypothesized mechanism of magnetosome formation

In particular, *M. magneticum* AMB-1 and *M. gryphiswaldense* MSR-1 are well investigated toward the elucidation of magnetosome formation because of the development of some important genetic engineering techniques for transformation (Okamura et al. 2003; Schultheiss and Schüler 2003), transposon insertion at random position in their genome (Matsunaga et al. 1992; Wahyudi et al. 2001; Li et al. 2005), specific gene deletion (Komeili and Vali 2004; Ullrich and Schüler 2010), and induction of target gene expression (Yoshino et al. 2010; Borg et al. 2014; Yamagishi et al. 2016). Each genetic engineering technique has provided us the chance of *in vivo* experiment in order to investigate the magnetosome formation. It notes that magnetotactic bacteria is an ideal model organism for *in vivo* investigation of biomineralization mechanism (especially nano-sized single crystal), because many biomineral synthesizing organisms, e.g., hydroxyapatite in the bones and teeth (Dorozhkin and Epple 2002), silica in diatoms (Kröger et al. 1999) and sponges (Shimizu et al. 1998), calcium carbonate in mollusk shells (Weiner and Addadi 1997), are eukaryote that is generally difficult to develop the genetic engineering techniques.

This section describes the overview of magnetosome formation mechanism in magnetotactic bacteria (Fig. 2.3). The mechanism has been analyzed based on the researches of key molecules by using the genetic engineering techniques. As current several studies pointed out the presence of different phenotypes between same gene lacking mutant strains of *M. magneticum* AMB-1 and *M. gryphiswaldense* MSR-1 (Murat et al. 2010; Raschdorf et al. 2016), herein we describe the researches mainly about *M. magneticum* AMB-1 to disturb the confusion. Firstly, a gene list reported to contribute to magnetosome formation was shown in Table 2.2 with the information of protein identification by proteomic studies in *M. magneticum* AMB-1 and *M. gryphiswaldense* MSR-1 (Okamura et al. 2000; Arakaki et al. 2003; Grünberg et al. 2004; Tanaka et al. 2006, 2010a). As found in this table, the magnetosome biogenesis occurred through multiple steps including magnetosome membrane formation by invagination of cytoplasmic membrane, magnetosome chain formation, iron transportation and redox control within magnetosome, magnetite crystallization, and morphological regulation. Hence, the researches in each step are described step by step, and the mechanism of magnetite crystallization and morphological regulation is briefly summarized because the content is shown in detail in Sect. 2.5.

2.3.1 Magnetosome Membrane Formation by Invagination of Cytoplasmic Membrane

A key step in the magnetosome organelle biogenesis is the magnetosome membrane formation by the invagination of cytoplasmic membrane. The compartmentalized nano-spaces provide domains for the regulation of iron concentrations and redox conditions, enabling the synthesis of morphologically controlled magnetite crystal in organisms. One of the most abundant proteins in magnetosome membrane, Mms16, was identified in *M. magneticum* AMB-1 (Okamura et al. 2000, 2001). As the Mms16 protein exhibited GTPase activity, which is known as an important activity for vesicle formation (e.g., endocytosis) in eukaryotic cells (Stenmark 2009; Colombo et al. 2014), the protein is suggested to be involved in the magnetosome membrane formation from cytoplasmic membrane. Within the proteomic study of magnetosome proteins in *M. magneticum* AMB-1, MpsA and the beta subunit, MpsB, were identified (Matsunaga et al. 2000; Tanaka et al. 2006). Given the homology between the MpsA and MpsB proteins with the *E. coli* acyl-CoA carboxylase and the supporting reports on the acylation and budding of vesicles (Randazzo et al. 1993; Chun et al. 2008), MpsA and MpsB are suggested to cooperate and mediate invagination of the cytoplasmic membrane to form magnetosome membrane.

One of the other dominant proteins in magnetosome membrane, MamA, has tetratricopeptide (TPR) repeat domain, which plays an important role in protein-protein interaction. The gene deletion in *M. magneticum* AMB-1 showed the decrease of magnetosome number, and the GFP fusions behave dynamic localization during cell growth. These analyses suggested that MamA assembling complex acts as a scaffold protein for magnetosome formation by interacting with other magnetosome proteins (Komeili and Vali 2004; Zeytuni et al. 2011; Nguyen et al. 2016).

The mutant strains of *M. magneticum* AMB-1 lacking any of *mamI*, *mamL*, *mamB*, and *mamQ* genes revealed the non-formation of magnetosome membrane (Murat et al. 2010). Therefore though these four proteins are suggested to be essential for magnetosome membrane formation, the precise molecular mechanism is largely unclear thus far. As MamL and MamI proteins are no homology with any proteins found in other bacteria, these proteins seem to have unique function in magnetotactic bacteria. In MamL, the C-terminal region mainly comprised by positively charged amino acids has been suggested to contribute for the formation of curved membrane structure through the interaction onto negatively charged phos-

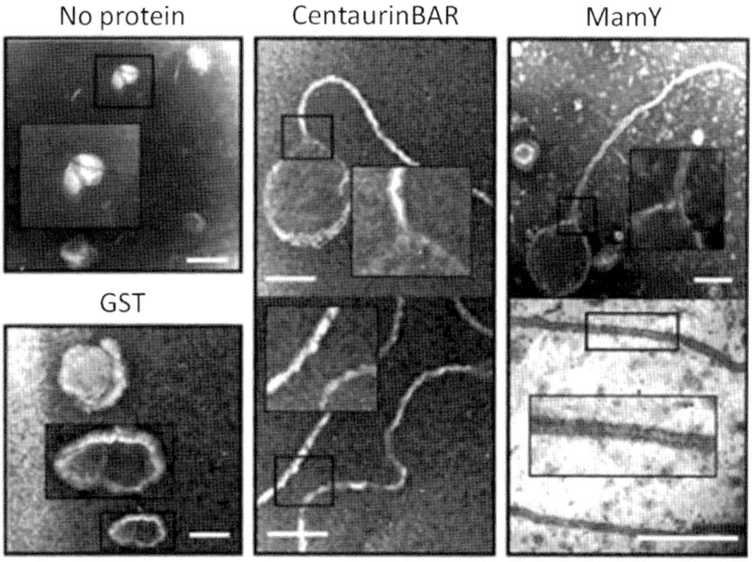

Fig. 2.4 Liposome tubulation assay with MamY protein. Electron micrographs of the liver lipid liposomes incubated with the proteins. MamY protein interacted with the liposome surfaces, leading to the formation of tubules with outer diameters of 20 nm. CentaurinBAR and GST proteins were used as the positive and negative controls, respectively. The inset shows liposomes at a higher magnification. Scale bar = 200 nm

pholipid surface, but further research may need to clarify the molecular function (Murat et al. 2010; Lohße et al. 2014; Raschdorf et al. 2016).

In order to form the intracellular spherical membranous structure (highly curved membrane structure) from relatively flat membrane structure (e.g., cell membrane), membrane curvature formation and constriction are important processes. In eukaryotic cells, for example, many endocytic proteins, such as dynamin and the bin/amphiphysin/Rvs (BAR) protein family, play a critical role during clathrin-mediated endocytosis (Peter et al. 2004; Shimada et al. 2007). On the other hand, in prokaryotes, methyl-accepting chemotaxis protein (MCP) in *E. coli* has been reported to deform the cell membrane structure by the protein overexpression (Khursigara et al. 2008). One potential protein attribute for the process in magnetosome membrane formation, MamY, was identified from comparative proteomic study (Tanaka et al. 2010a). The protein was only present in magnetosomes containing small immature crystals, whereas it was not identified in magnetosome containing regular mature crystals or any other cellular fraction. The protein has weak homology to eukaryotic BAR protein with alpha helix repeats from secondary structure prediction. *In vivo* study using *mamY* gene deletion mutant showed the formation of bigger magnetosome vesicles with a larger number of small magnetite crystals. The recombinant protein produced in *E. coli* revealed unique characteristics of direct lipid binding and liposome tubulation (Fig. 2.4). In addition, from the aspect of 3D structure prediction, the protein showed low conservation to the domain from Talin protein, which is a cytoskeletal linker protein connecting between actin and membrane structure (Niggli et al. 1994; Nudelman and Zarivach 2014). Based on these researches, the function of MamY protein is considered to be involving in the constriction/deformation of cytoplasmic membrane in order to form magnetosome membrane or size control of magnetosome membranes.

2.3.2 Magnetosome Chain Formation

Magnetosomes intracellularly align in linear structure along the major axis of cell in *M. magneticum* AMB-1. The chain structure seems to be mainly controlled by two proteins, MamK and MamJ. In both *mamK* and *mamJ* deletion mutants of magnetotactic bacteria, mutant cells lack the ability of magnetosome chain formation (Scheffel et al. 2006; Komeili et al. 2006; Draper et al. 2011). MamK is a homologue of the bacterial actin-like protein MreB and is revealed to form filament structures (Komeili et al. 2006). MamK nucleates at multiple sites to form long filaments that were confirmed via protein expression in *E. coli* (Pradel et al. 2006). *In vitro* polymerization of recombinant MamK was examined, and formation of long filamentous bundles was observed (Taoka et al. 2007). MamJ, an acidic protein, interacts with the filamentous MamK protein structure (Pan et al. 2012), and the *mamJ* deletion mutant showed the lack of magnetosome chain structure (Scheffel et al. 2006). From these observations, it is assumed that MamJ is involved in the assembly of magnetosomes into chains by connecting them to the filamentous structure formed by MamK (Scheffel and Schüler 2007).

2.3.3 Iron Transportation and Redox Control Within Magnetosome

M. magneticum AMB-1 can accumulate 100 times higher concentration of iron rather than *E. coli* (Nakamura et al. 1995; Suzuki et al. 2006). Therefore, effective biological system for iron uptake should be facilitated in the magnetosome formation. The *magA*, the gene encoding a proton-driving H^+/Fe (II) antiporter protein, was firstly isolated as an iron transporting protein candidate related to magnetosome formation (Nakamura et al. 1995). The gene was screened from non-magnetosome forming mutant library through the Tn5 transposon-based treatment. As iron accumulation ability of liposomes produced from MagA expressing *E. coli* was demonstrated, the protein is suggested to be an iron transporter in *M. magneticum* AMB-1. Other candidates of iron transporters, MamB and MamM, share 47% sequence similarity. The proteins are homologue of cation diffusion facilitator (CDF) protein family. MamM has been relatively well investigated, and the deletion or single amino acid mutations coursed severe effect for magnetite crystal formation including complete inhibition of crystallization (Murat et al. 2010; Uebe et al. 2011; Zeytuni et al. 2014).

M. magneticum AMB-1 produces magnetite, Fe_3O_4 comprising of two Fe(III) and one Fe(II). As iron ion is generally uptaking Fe(II) within the cell, the redox condition is believed to be highly controlled to produce the particles. An aldehyde ferredoxin oxidoreductase (AOR), which plays a role in aldehyde oxidation, was also isolated from transposon mutant library. Iron uptake and growth of AOR-deficient mutant were lower than wild type under microaerobic condition (Wahyudi et al. 2003). The results indicate that AOR may contribute to ferric iron reduction during magnetosome synthesis under microaerobic respiration. In addition, mutation in MamP causes deficiencies in crystal maturation and the loss of response for external magnetic moment (Siponen et al. 2013; Jones et al. 2015). The protein possesses CXXCH motif of typical c-type cytochrome heme-binding motif. MamP contributes for the oxidation of iron to form iron(III) ferrihydrite (Murat et al. 2010). In addition to MamP, the motif was shared in several magnetosome proteins including MamE, MamT, and MamX. As these deletions caused the production of hematite or extremely smaller crystals rather than large magnetite crystal in wild type (Yang et al. 2013; Raschdorf et al. 2013), these proteins also seem to have important role for iron-redox control during magnetosome formation.

2.3.4 The Mechanism of Magnetite Crystallization and Morphological Regulation

During long history of research about functional protein analyses in magnetotactic bacteria, a series of Mms proteins comprising Mms5, 6, 7, and 13 were firstly identified as candidate proteins directly involved in crystal formation of biogenic magnetic nano-particles (Arakaki et al. 2003). Through *in vivo* and *in vitro* analyses, these proteins have disclosed the fact that the bacteria regulate the size and morphology of nano-sized biomineral using biological molecules in ambient condition in bacterial cells (see Sect. 2.5). In addition to these Mms proteins, several proteins have been also suggested to involve in the process. For example, MmsF protein is a protein encoded within *mms6* operon in *M. magneticum* AMB-1, and the gene deletion mutant strain synthesized smaller crystals and lower magnetic responses (Murat and Falahati 2012; Rawlings et al. 2014). The mutant strain lacking *mamG* gene revealed the phenotype of spherical magnetite crystal formation with smaller sizes. In *M. gryphiswaldense* MSR-1, *mamF* deletion caused no effect for magnetite nanoparticle numbers and sizes, but the additional deletion of *mmsF* revealed further noticeable phenotypic change rather than only *mmsF* gene deletion mutant (Lohße et al. 2014). Taken together, magnetotactic bacteria seem to achieve the biological unique process about magnetic nano-particle regulation using a group of Mms proteins and several other proteins.

2.4 Nano-particle Encapsulating Lipid Tubule Synthesis Inspired from Prokaryotic Membrane Deformation Protein, MamY

As described in above section, magnetotactic bacterial possesses complex biological process for intracellular membranous structure formation through the invagination of cytoplasmic membrane. Among various proteins suggested to have contribute for the process, MamY possesses a domain with slight homology (Tanaka et al. 2010a) to the eukaryotic bin/amphiphysin/Rvs167 (BAR) domain (Peter et al. 2004; Shimada et al. 2007), a region commonly found in liposome tubulation proteins, and it has been experimentally proven that MamY does show both activities of direct binding to lipid and of liposome tubulation *in vitro*, similar to that of its eukaryotic counterparts (Tanaka et al. 2010a). It is proposed that this liposome deformation mechanism by BAR protein is caused by the banana-shaped protein domain binding to the membrane (by mainly electrostatic interaction) and self-assembling to form a spiral-like structure surrounding the tubulated membrane (Peter et al. 2004; Shimada et al. 2007). This biological phenomenon has inspired the development of novel technology to induce the 1D assembly of nano-particles within the lipid bilayer in ambient conditions.

To date, liposomes, one of the most recognized platforms for crystalline formation in size regulation of nano-materials, have been widely introduced in various applications due to their biocompatibility and their superior dispersion properties (due to the electrostatic charges of phospholipids). However, it is commonly observed and understood that the morphology of various nano-materials synthesized within liposome platforms is strictly limited to the synthesis of materials of circular structures (Fig. 2.5). This setback has thus brought about much limitation to the application such as metal nano-wires within micro-fabricated devices. However, if we look at organisms including magnetotactic bacteria, various nano- and micro-membranous structures made by lipid and protein can be seen including spherical, tubular, and lamella in a cell. In order to properly function, all these structures seem to be strictly controlled by biological molecules. Then, if we could elucidate and

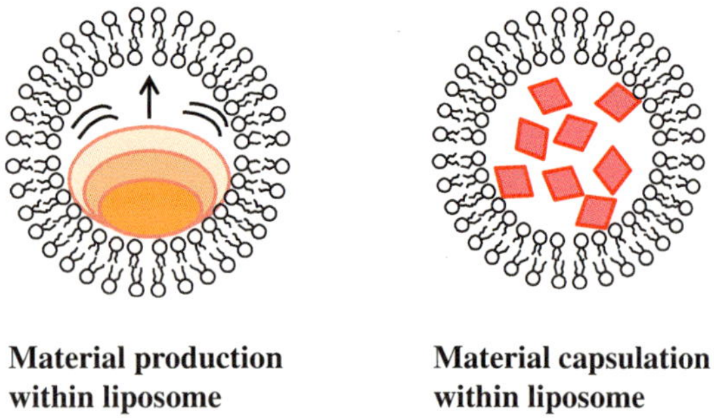

Fig. 2.5 Images of material production capsulated in lipid bilayer membrane through two different courses. One is material production within liposome, and the other is material capsulation

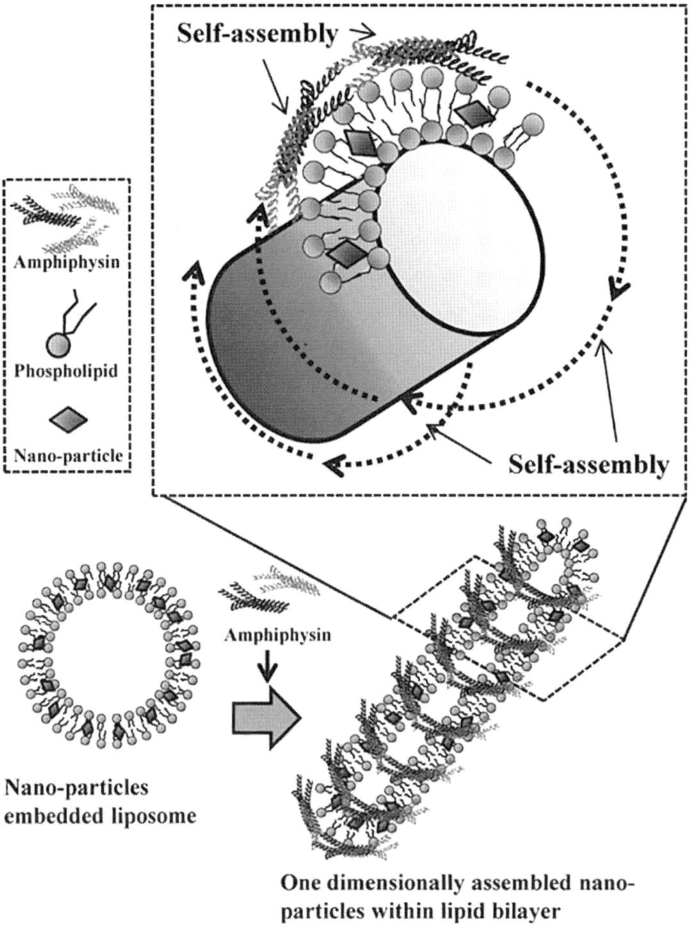

Fig. 2.6 Schematic image of the fabrication of tubulated liposomes containing NPs within the lipid bilayer

mimic the machinery, novel structure controlled membrane-based materials can be fabricated. Therefore, based on the inspiration from MamY functional research, we tried to fabricate new nano-material using BAR protein, which shows structure and function similarities with MamY protein.

The fabrication process consists of two simple steps (Tanaka et al. 2012b) (Fig. 2.6). Firstly, giant unilamellar vesicles (GUVs) containing hydrophobic nanoparticles were formed by a rehydration method. Secondly, liposome tubulation was performed using BAR protein (Amphiphysin). Based on this simple strategy, one-dimensional membrane structures (lipid tubule) containing various nano-particles ($CuInS_2$ and CdSe quantum dots and gold) were successfully fabricated (Fig. 2.7). By choosing various NPs, the lipid tubules could be designed to have a range of properties. Additionally, the optically observable liposome/lipid tubule (low-/high-

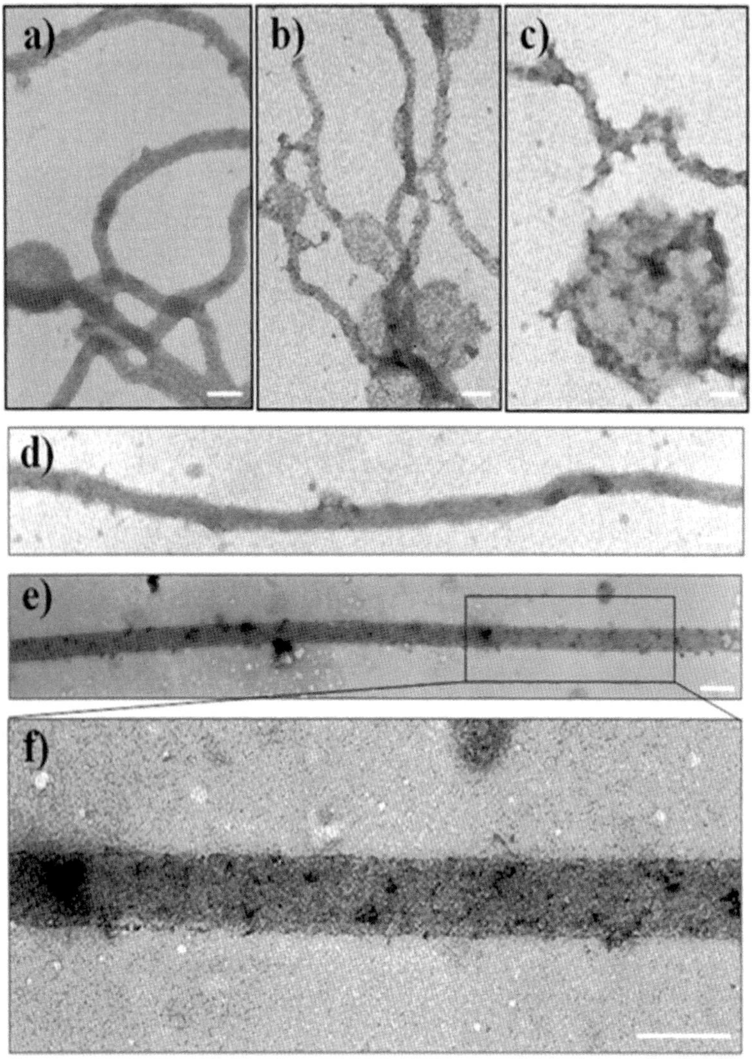

Fig. 2.7 TEM images of tubulated liposome containing quantum dots within the lipid bilayer: (**a**) without QDs; (**b**, **c**) with QDs (QD:lipid = 1:5000 and 1:1000) using Amphiphysin-BAR protein. (**d**–**f**) Representative images of tubulated liposome (**d**) without QDs, (**e**) with QDs (QD:lipid = 1:5000), and (**f**) magnified image of the area indicated in (**e**). Scale bars indicate 50 nm

curvature membrane) with high photostability could be a useful tool for elucidation of biological reactions onto/within highly curved membranous structures. The developed material with unique characteristics, such as protein-induced morphological change, biocompatibility, high dispersity, and the potential of diameter tuning with a further molecular engineering technique, could be useful for various applications in nanobiotechnology.

2.5 Crystal Size and Shape Determination Mechanism of Magnetic Nano-particles

Finely tuned size and shape of biominerals are directed by the functions of specific proteins in biominerals, although the knowledge of the mechanisms is still largely limited so far. To clarify this, isolation of proteins involved in the formation of magnetite crystal has been examined in *M. magneticum* AMB-1 (Arakaki et al. 2003). As described above, strain AMB-1 produces cubo-octahedral-shaped particles mainly consisting of {111} and {100} crystal faces. Since such proteins are considered to have function near the biomineral crystals, they should locate inside of biominerals or closely associated with the surface. As an approach to isolate such biomineralization proteins, proteins tightly attaching to the mineral surface were extracted to use surfactant (1% SDS) with boiling treatment (100 °C). As a result, a group of protein designated as Mms5, Mms6, Mms7, and Mms13 were successfully isolated. They showed similarities in the N-terminal and C-terminal amino acid sequence and were speculated to have related functions in the magnetic nano-particle formation (Arakaki et al. 2003).

Protein functions are generally analyzed from different ways such as *in vivo*, *in vitro*, and *in silico*. Each approach has strength and weakness for determination of protein functions, and thus it is best to use some of these to complement obtained results. For example, *in vivo* analysis, such as mutagenesis, is suitable to show the protein function in the cellular system. In contrast, *in vitro* analysis may be able to clarify the function and structural information at molecular level. *In silico* analysis is useful to reinforce the experimental results in *in vivo* and *in vitro*. Regarding to *in vivo* analysis of target protein, gene deletion technique is a powerful tool in order to understand the protein function in the cellular system. To elucidate *in vivo* function of Mms6 protein, the *mms6* gene deletion mutant of *M. magneticum* AMB-1 was constructed by homologous recombination (Tanaka et al. 2011b). When the *mms6* gene was deleted from the genome of magnetotactic bacteria, the gene deletion mutant synthesized small particles in their cells (Tanaka et al. 2011b). The average particle size (27.4 ± 8.9 nm) and shape factor (0.74 ± 0.23 nm) for the gene deletion mutant were significantly different from the wild type, with 48.3 ± 12.5 nm particle size and 0.92 ± 0.16 nm shape factor (Fig. 2.8). The crystals formed in the gene deletion mutant exhibited high index crystal faces, while the wild-type strain synthesized cubo-octahedral crystals with {111} and {100} faces. However, the number of magnetic particles in the mutant remained unaffected. These observations indicate that Mms6 does not function in the nucleation step of magnetite biomineralization (Tanaka et al. 2011b). Similarly, gene deletion mutagenesis was also conducted for other *mms* genes (Arakaki et al. 2014). The *mms7* gene deletion mutant produced elongated crystals as those in the *mms6* gene deletion mutant (Fig. 2.8). In contrast, the gene deletion mutants of *mms5* or *mms7* produced cubo-octahedral crystals with slightly smaller average sizes than those in the wild-type strain (Arakaki et al. 2014). Thus, Mms proteins are responsible for crystal growth and controlling the geometries of the crystal surface structures,

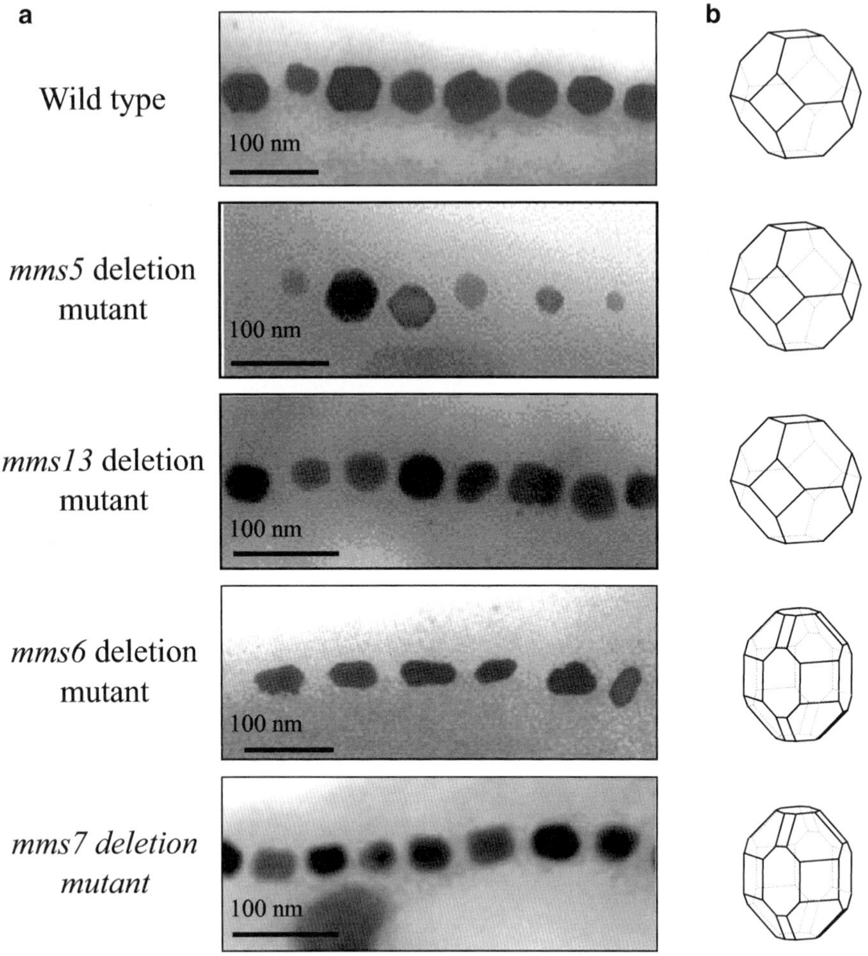

Fig. 2.8 (**a**) Transmission electron micrograph of magnetic nano-particles formed within various *mms* gene deletion mutant strains. Scale bar, 50 nm. (**b**) Ideal models of the particles

although they have some differences in the function (Fig. 2.9). Moreover, other magnetotactic bacterial strains that form bullet-shaped and/or pleomorphic magnetite crystals do not possess *mms* genes (Nakazawa and Arakaki 2009). The genes are specific for the strains which produce cubo-octahedral magnetic particles. The specified and cooperative functions of these proteins resulted in the characteristic cubo-octahedral crystal morphology in strain AMB-1.

The mutagenesis studies revealed that the expression of Mms proteins alters the size and shape of magnetic nano-particles in magnetotactic bacteria. While preparing the gene mutant strains, a spontaneous mutant strain that produced novel dumbbell-shaped magnetic particles was obtained (Arakaki et al. 2014). Interestingly, the crystal shapes cannot be produced by chemical synthetic pathways in artificial

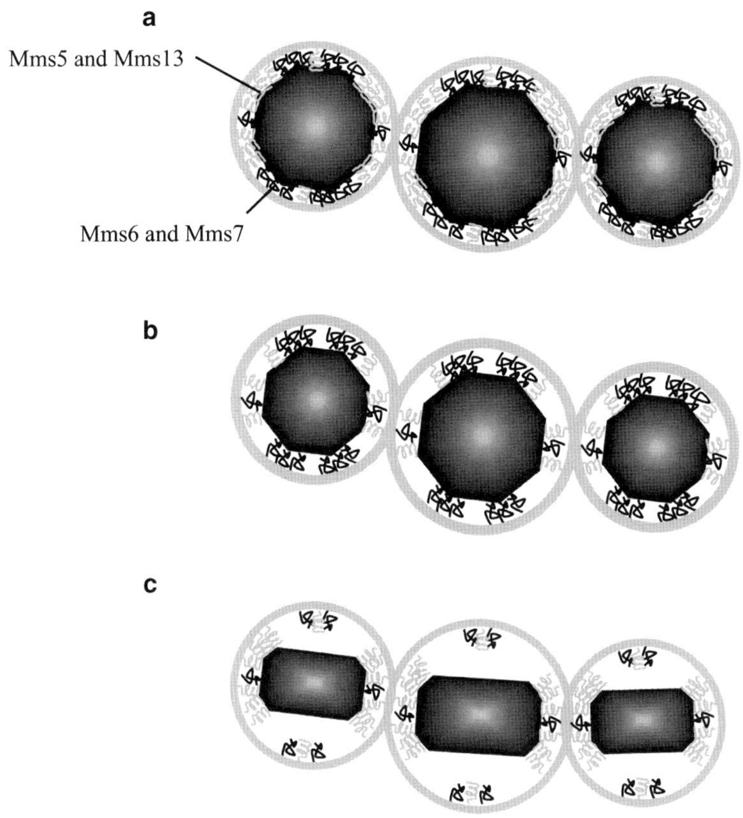

Fig. 2.9 Morphological regulation mechanism of magnetic nano-particle by the coordinated action of Mms proteins. Model of magnetic nano-particles formed in the wild type (**a**), *mms5* or *mms13* gene deletion mutant (**b**), and *mms6* or *mms7* gene deletion mutant (**c**). Localization of the Mms proteins in magnetosome vesicle defines the crystal morphology of nano-particle

systems. These findings imply that the crystal morphology can be controlled via regulation of the expression of genes encoding Mms proteins in a cell system. The attempt was conducted to construct inducible *mms7* gene expression system in the mutant strain producing dumbbell-shaped crystals (Yamagishi et al. 2016). A tetracycline-inducible expression system was used for regulation of *mms7* gene expression. Increasing concentration of the inducer molecule (anhydrotetracycline: ATc) in the media triggers expression levels of Mms7 protein in bacterial cells, and the expressed protein promotes the crystal growth to certain direction. As a result, crystal shape of magnetic particles successfully regulated from dumbbell to cubo-octahedron (Fig. 2.10). According to the shape change, average crystal size was increased from 23 to 29 nm (Yamagishi et al. 2016). The regulation of magnetite biomineralization provides new opportunities for the *in vivo* synthesis of magnetite materials with tuned size, shape, and properties. The preparation method will expand the biotechnological applications of the magnetic nano-particles.

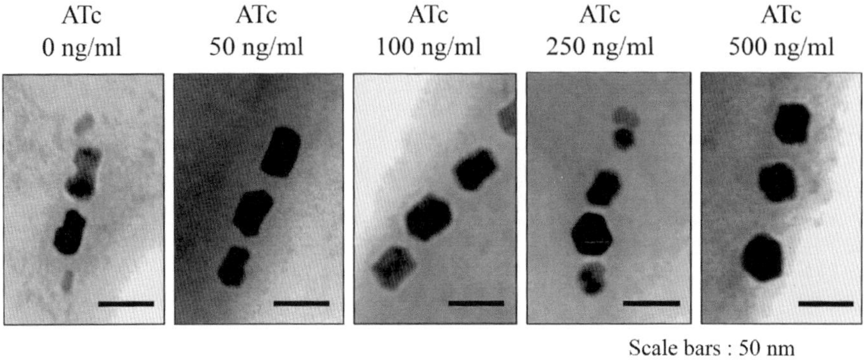

Fig. 2.10 Control of particle size and shape by regulation of *mms7* gene using a tetracycline-inducible expression system. Magnetite crystals synthesized in a spontaneous mutant strain that produces dumbbell-shaped magnetic particles. A plasmid containing *mms7* gene with the tetracycline-inducible promoter was introduced in the mutant strain. The transformants were cultivated in the presence of 50–500 ng/mL anhydrotetracycline (ATc)

2.6 Crystal and Granule Formation from Toxic or Useful Compounds Through the Reduction Bioprocess Within Magnetotactic Bacteria

An important research area in nano-material science deals with the nano-particle synthesis showing various properties of different morphologies and sizes. The synthesis of inorganic nano-materials and nanostructures by the means of chemically, physically, and biologically controlled courses has been developed. Among these, biologically controlled biomaterial synthesis has attracted various scientists due to the potential for the eco-friendly process development. Especially in bacteria, a wide number of bacterial species including magnetotactic bacteria have been used in green nanotechnology for nano-particle production. Microorganisms have acquired various strategies (metabolic pathways) toward taking advantage of any ecological niches found in environment, in order to survive in the most widely varied environmental conditions on Earth. Some organisms acquired the tolerance for toxic molecule, and some are able to convert various toxic targets to relatively non-toxic molecules. For example, gold, silver, platinum, palladium, titanium, cadmium sulfide, magnetite, and so on have been reported to be synthesized using various bacterial species (Barton et al. 2015; Sharma et al. 2015).

In regard to magnetotactic bacteria, they are unique prokaryotes, recognized by their response to a magnetic field because of the presence of magnetic nano-particles of magnetite or greigite within the cells. The magnetic nano-particles confer a magnetic moment to the cells, allowing them to migrate in aquatic environments under the influence of the Earth's geomagnetic field. Such unique characteristic allows the bacteria to be applied not only for nano-particle production but also for bioremediation and can be easily collected and concentrated by magnetic separation.

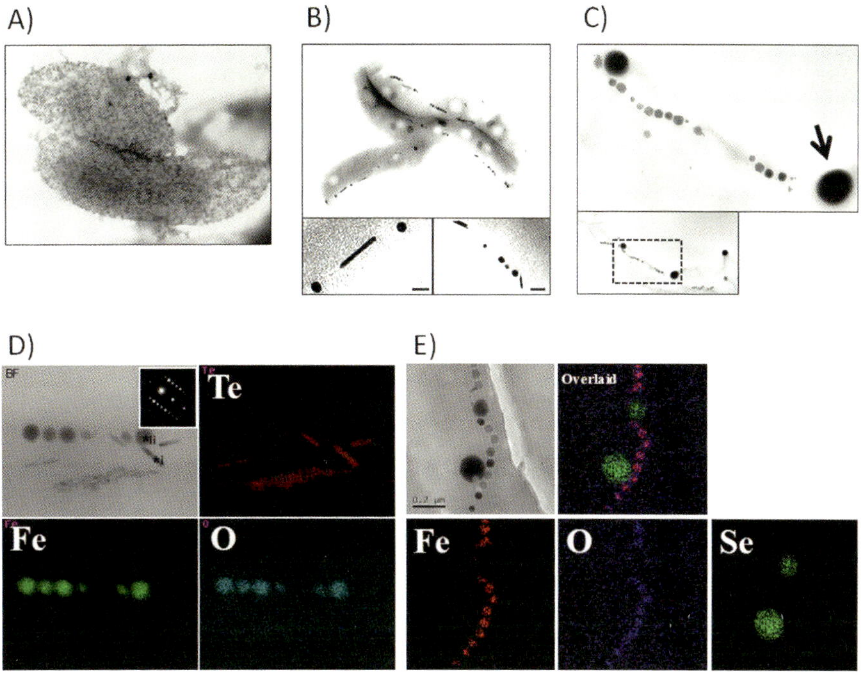

Fig. 2.11 Scheme of various nano-material productions on and within magnetotactic bacteria. Transmission electron micrographs of *D. magneticus* strain RS-1 grown in the presence of cadmium (**a**) and *M. magneticum* strain AMB-1 grown in the presence of tellurite (**b**) and selenite (**c**). STEM-EDX maps of Se, TE, Fe, and O in *M. magneticum* strain AMB-1 grown in the presence of tellurite (**d**) and selenite (**e**)

Environmental remediation, a technique of waste removal and/or neutralization of pollutant from a contaminated site, is an attractive field because of the increasing difficulty and importance of pure water acquisition in both developing and industrial countries. Hence, studies of the target biosorption and accumulation onto/within microorganisms have been well demonstrated, but cell recovery remains a bottleneck in this approach because scale-up of collection methods, such as centrifugation and filtration, provides a huge logistical and monetary challenge.

Using magnetotactic bacteria, studies for removal of toxic molecules were conducted in 1991 (Bahaj et al. 1991). From then, various target molecules including heavy metals and trace radionuclides were tried to remove using this organism (Bahaj et al. 1994, 1998a, b). Subsequently, the adsorption and mineralization of metal ions were investigated (Arakaki et al. 2002; Keim et al. 2005; Huiping et al. 2007; Song et al. 2008; Tanaka et al. 2008, 2010b, 2011a, 2016; Cai et al. 2011). We have investigated the use of magnetic bacteria (*D. magneticus* RS-1 and *M. magneticum* AMB-1) for nano-particle productions from toxic and/or valuable metals and metalloid such as Cd^{2+} (Arakaki et al. 2002;), Te (Tanaka et al. 2010b), and Se (Tanaka et al. 2016) to CdS, elemental Te particles, and Se granules (Fig. 2.11).

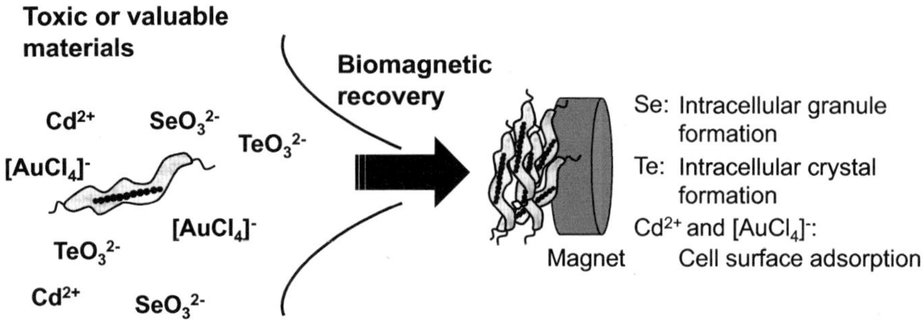

Fig. 2.12 Schematic view of biomagnetic recovery of toxic and/or valuable molecules using magnetotactic bacteria

Importantly, the Se- and Te-accumulating bacteria were successfully recovered with an external magnetic field. Therefore, we believe that magnetotactic bacteria have the unique advantage for cell recovery, providing a new effective methodology for bioremediation of water environment and for further nano-material production from pollutant product (Fig. 2.12).

It also notes that *M. magneticum* AMB-1 seems to be not metallophile (metal resistance bacteria) through the profiling of various trace metal tolerance (Tanaka et al. 2012a, 2016). Current genetic and environmental microbiological research shows that magnetic nano-particle production within bacteria occurs across a diverse group of bacterial species. In fact, the genetic region corresponding to magnetosome formation, called the magnetosome island (MAI), is found within microbes spread across the phylogenetic tree. Therefore, a magnetotactic bacteria species with higher tolerance and effective accumulation of target molecules could be found and used to improve the nano-particle production and the bioremediation. In addition, recently, magnetosome formation was enabled in another bacterial species by artificially transferring key genetic regions of the MAI into the host organism (Kolinko et al. 2014). Therefore, the induction of magnetosome formation within known bacteria showing high resistance to a target element and various material bioproduction is another promising approach for various applications.

2.7 Summary and Perspective

Understandings of magnetosome formation mechanism in magnetotactic bacteria have tremendously improved by the recent comprehensive molecular studies. Numbers of functional molecules involved in the magnetosome formation have been identified. The molecules and the cellular processes understood from the fundamental studies will allow us to develop various magnetic nano-materials. The morphologically controlled magnetic nano-particles generated by both chemical processes and genetic manipulations are useful for various applications in biotechnology and magnetic recordings. Advanced magnetic materials through environmentally benign routes will also be created for a wide array of tailor-made applications via collaborative works in a wide variety of scientific fields, including inorganic chemistry, organic chemistry, biology, computer science, and engineering.

References

Abreu F, Morillo V, Nascimento FF et al (2014) Deciphering unusual uncultured magnetotactic multicellular prokaryotes through genomics. ISME J 8:1055–1068. https://doi.org/10.1038/ismej.2013.203

Arakaki A, Takeyama H, Tanaka T, Matsunaga T (2002) Cadmium recovery by a sulfate-reducing magnetotactic bacterium, *Desulfovibrio magneticus* RS-1, using magnetic separation. Appl Biochem Biotechnol 98–100:833–840. https://doi.org/10.1385/ABAB:98-100:1-9:833

Arakaki A, Webb J, Matsunaga T (2003) A novel protein tightly bound to bacterial magnetic particles in *Magnetospirillum magneticum* strain AMB-1. J Biol Chem 278:8745–8750. https://doi.org/10.1074/jbc.M211729200

Arakaki A, Nakazawa H, Nemoto M et al (2008) Formation of magnetite by bacteria and its application. J R Soc Interface 5:977–999. https://doi.org/10.1098/rsif.2008.0170

Arakaki A, Shibusawa M, Hosokawa M, Matsunaga T (2010) Preparation of genomic DNA from a single species of uncultured magnetotactic bacterium by multiple-displacement amplification. Appl Environ Microbiol 76:1480–1485. https://doi.org/10.1128/AEM.02124-09

Arakaki A, Yamagishi A, Fukuyo A et al (2014) Co-ordinated functions of Mms proteins define the surface structure of cubo-octahedral magnetite crystals in magnetotactic bacteria. Mol Microbiol 93:554–567. https://doi.org/10.1111/mmi.12683

Bahaj AS, Ellwood DC, Watson JHP (1991) Extraction of heavy metals using microorganisms and high gradient magnetic separation. Solutions 27:6–9

Bahaj AS, James PAB, Croudace IW (1994) Metal uptake and separation using magnetotactic bacteria. IEEE Trans Magn 30:4707–4709

Bahaj AS, Croudace IW, James PAB et al (1998a) Continuous radionuclide recovery from wastewater using magnetotactic bacteria. J Magn Magn Mater 184:241–244. https://doi.org/10.1016/S0304-8853(97)01130-X

Bahaj AS, James PAB, Moeschler FD (1998b) Low magnetic-field separation system for metal-loaded magnetotactic bacteria. J Magn Magn Mater 177–181:1453–1454. https://doi.org/10.1016/S0304-8853(97)00677-X

Barton LL, Tomei-Torres FA, Xu H, Zocco T (2015) Metabolism of metals and metalloids by the sulfate-reducing bacteria. In: Bacteria-metal interactions. Springer International Publishing, Cham, pp 57–83

Bazylinski DA, Frankel RB, Heywood BR et al (1995) Controlled biomineralization of magnetite (Fe(inf3)O(inf4)) and greigite (Fe(inf3)S(inf4)) in a magnetotactic bacterium. Appl Environ Microbiol 61:3232–3239

Bazylinski DA, Williams TJ, Lefèvre CT et al (2013) *Magnetovibrio blakemorei* gen. nov., sp. nov., a magnetotactic bacterium (Alphaproteobacteria: Rhodospirillaceae) isolated from a salt marsh. Int J Syst Evol Microbiol 63:1824–1833. https://doi.org/10.1099/ijs.0.044453-0

Bergeron JRC, Hutto R, Ozyamak E et al (2016) Structure of the magnetosome-associated actin-like MamK filament at sub-nanometer resolution. Protein Sci. 26:93–102. https://doi.org/10.1002/pro.2979

Blakemore R (1975) Magnetotactic bacteria. Science 190:377–379

Borg S, Hofmann J, Pollithy A et al (2014) New vectors for chromosomal integration enable high-level constitutive or inducible magnetosome expression of fusion proteins in *Magnetospirillum gryphiswaldense*. Appl Environ Microbiol 80:2609–2616. https://doi.org/10.1128/AEM.00192-14

Cai F, Li J, Sun J, Ji Y (2011) Biosynthesis of gold nanoparticles by biosorption using *Magnetospirillum gryphiswaldense* MSR-1. Chem Eng J 175:70–75. https://doi.org/10.1016/j.cej.2011.09.041

Chun J, Shapovalova Z, Dejgaard SY et al (2008) Characterization of class I and II ADP-ribosylation factors (Arfs) in live cells: GDP-bound class II Arfs associate with the ER-Golgi intermediate compartment independently of GBF1. Mol Biol Cell 19:3488–3500. https://doi.org/10.1091/mbc.E08-04-0373

Colombo M, Raposo G, Théry C (2014) Biogenesis, secretion, and intercellular interactions of exosomes and other extracellular vesicles. Annu Rev Cell Dev Biol 30:255–289. https://doi.org/10.1146/annurev-cellbio-101512-122326

Ding Y, Li J, Liu J et al (2010) Deletion of the *ftsZ*-like gene results in the production of superparamagnetic magnetite magnetosomes in *Magnetospirillum gryphiswaldense*. J Bacteriol 192:1097–1105

Dorozhkin SV, Epple M (2002) Biological and medical significance of calcium phosphates. Angew Chem Int Ed Eng 41:3130–3146. https://doi.org/10.1002/1521-3773(20020902)41:17<3130::AID-ANIE3130>3.0.CO;2-1

Draper O, Byrne ME, Li Z et al (2011) MamK, a bacterial actin, forms dynamic filaments *in vivo* that are regulated by the acidic proteins MamJ and LimJ. Mol Microbiol 82:342–354. https://doi.org/10.1111/j.1365-2958.2011.07815.x

Fukuda Y, Okamura Y, Takeyama H, Matsunaga T (2006) Dynamic analysis of a genomic island in *Magnetospirillum* sp. strain AMB-1 reveals how magnetosome synthesis developed. FEBS Lett 580:801–812. https://doi.org/10.1016/j.febslet.2006.01.003

Gorby YA, Beveridge TJ, Blakemore RP (1988) Characterization of the bacterial magnetosome membrane. J Bacteriol 170:834–841

Grünberg K, Müller E, Otto A (2004) Biochemical and proteomic analysis of the magnetosome membrane in *Magnetospirillum gryphiswaldense*. Appl Environ Microbiol 70:1040–1050

Guo FF, Yang W, Jiang W et al (2012) Magnetosomes eliminate intracellular reactive oxygen species in *Magnetospirillum gryphiswaldense* MSR-1. Environ Microbiol 14:1722–1729. https://doi.org/10.1111/j.1462-2920.2012.02707.x

Hershey DM, Ren X, Melnyk RA et al (2016) MamO is a repurposed serine protease that promotes magnetite biomineralization through direct transition metal binding in magnetotactic bacteria. PLoS Biol. https://doi.org/10.1371/journal.pbio.1002402

Huiping S, Xingang L, Jinsheng S et al (2007) Biosorption equilibrium and kinetics of Au (III) and Cu (II) on. Chin J Chem Eng 15:847–854

Ji B, Zhang S-D, Arnoux P et al (2014) Comparative genomic analysis provides insights into the evolution and niche adaptation of marine *Magnetospira* sp. QH-2 strain. Environ Microbiol 16:525–544. https://doi.org/10.1111/1462-2920.12180

Jogler C, Wanner G, Kolinko S et al (2011) Conservation of proteobacterial magnetosome genes and structures in an uncultivated member of the deep-branching *Nitrospira* phylum. Proc Natl Acad Sci 108:1134–1139. https://doi.org/10.1073/pnas.1012694108

Jones SR, Wilson TD, Brown ME et al (2015) Genetic and biochemical investigations of the role of MamP in redox control of iron biomineralization in *Magnetospirillum magneticum*. Proc Natl Acad Sci U S A 112:3904–3909. https://doi.org/10.1073/pnas.1417614112

Keim CN, Solórzano G, Farina M, Lins U (2005) Intracellular inclusions of uncultured magnetotactic bacteria. Int Microbiol 8:111–117. doi: im2305016

Khursigara CM, Wu X, Zhang P et al (2008) Role of HAMP domains in chemotaxis signaling by bacterial chemoreceptors. Proc Natl Acad Sci U S A 105:16555–16560. https://doi.org/10.1073/pnas.0806401105

Kolinko S, Jogler C, Katzmann E et al (2012) Single-cell analysis reveals a novel uncultivated magnetotactic bacterium within the candidate division OP3. Environ Microbiol 14:1709–1721. https://doi.org/10.1111/j.1462-2920.2011.02609.x

Kolinko I, Lohße A, Borg S et al (2014) Biosynthesis of magnetic nanostructures in a foreign organism by transfer of bacterial magnetosome gene clusters. Nat Nanotechnol 9:193–197. https://doi.org/10.1038/nnano.2014.13

Komeili A, Vali H (2004) Magnetosome vesicles are present before magnetite formation, and MamA is required for their activation. Proc Natl Acad Sci U S A 101:3839–3844

Komeili A, Li Z, Newman DK, Jensen GJ (2006) Magnetosomes are cell membrane invaginations organized by the actin-like protein MamK. Science 311:242–245. https://doi.org/10.1126/science.1123231

Koziaeva VV, Dziuba MV, Ivanov TM et al (2016) Draft genome sequences of two magnetotactic bacteria, *Magnetospirillum moscoviense* BB-1 and *Magnetospirillum marisnigri* SP-1. Genome Announc 4:e00814–e00816. https://doi.org/10.1128/genomeA.00814-16

Kröger N, Deutzmann R, Sumper M (1999) Polycationic peptides from diatom biosilica that direct silica nanosphere formation. Science 286:1129–1132

Lefèvre CT, Wu L-F (2013) Evolution of the bacterial organelle responsible for magnetotaxis. Trends Microbiol. https://doi.org/10.1016/j.tim.2013.07.005

Lefèvre CT, Trubitsyn D, Abreu F et al (2013) Comparative genomic analysis of magnetotactic bacteria from the Deltaproteobacteria provides new insights into magnetite and greigite magnetosome genes required for magnetotaxis. Environ Microbiol 15:2712–2735

Li F, Li Y, Jiang W et al (2005) Cloning and functional analysis of the sequences flanking mini-Tn5 in the magnetosomes deleted mutant NM4 of *Magnetospirillum gryphiswaldense* MSR-1. Sci China C Life Sci 48:574–584. https://doi.org/10.1360/062005-26

Lin W, Deng A, Wang Z et al (2014) Genomic insights into the uncultured genus "*Candidatus* Magnetobacterium" in the phylum *Nitrospirae*. ISME J 8:2463–2477. https://doi.org/10.1038/ismej.2014.94

Lohße A, Borg S, Raschdorf O et al (2014) Genetic dissection of the *mamAB* and *mms6* operons reveals a gene set essential for magnetosome biogenesis in *Magnetospirillum gryphiswaldense*. J Bacteriol 196:2658–2669. https://doi.org/10.1128/JB.01716-14

Matsunaga T, Kamiya S (1987) Use of magnetic particles isolated from magnetotactic bacteria for enzyme immobilization. Appl Microbiol Biotechnol 26:328–332. https://doi.org/10.1007/BF00256663

Matsunaga T, Nakamura C, Burgess JG, Sode K (1992) Gene transfer in magnetic bacteria: transposon mutagenesis and cloning of genomic DNA fragments required for magnetosome synthesis. J Bacteriol 174:2748–2753

Matsunaga T, Tsujimura N, Okamura Y, Takeyama H (2000) Cloning and characterization of a gene, *mpsA*, encoding a protein associated with intracellular magnetic particles from *Magnetospirillum* sp. strain AMB-1. Biochem Biophys Res Commun 268:932–937. https://doi.org/10.1006/bbrc.2000.2236

Matsunaga T, Okamura Y, Fukuda Y et al (2005) Complete genome sequence of the facultative anaerobic magnetotactic bacterium *Magnetospirillum* sp. strain AMB-1. DNA Res 12:157–166. https://doi.org/10.1093/dnares/dsi002

Matsunaga T, Nemoto M, Arakaki A, Tanaka M (2009) Proteomic analysis of irregular, bullet-shaped magnetosomes in the sulphate-reducing magnetotactic bacterium *Desulfovibrio magneticus* RS-1. Proteomics 9:3341–3352. https://doi.org/10.1002/pmic.200800881

Müller FD, Raschdorf O, Nudelman H et al (2014) The FtsZ-like protein FtsZm of *Magnetospirillum gryphiswaldense* likely interacts with its generic homolog and is required for biomineralization under nitrate deprivation. J Bacteriol 196:650–659. https://doi.org/10.1128/JB.00804-13

Murat D, Falahati V (2012) The magnetosome membrane protein, MmsF, is a major regulator of magnetite biomineralization in *Magnetospirillum magneticum* AMB-1. Mol Microbiol 85:684–699

Murat D, Quinlan A, Vali H, Komeili A (2010) Comprehensive genetic dissection of the magnetosome gene island reveals the step-wise assembly of a prokaryotic organelle. Proc Natl Acad Sci U S A 107:5593–5598

Nakamura N, Hashimoto K, Matsunaga T (1991) Immunoassay method for the determination of immunoglobulin G using bacterial magnetic particles. Anal Chem 63:268–272

Nakamura C, Burgess JG, Sode K, Matsunaga T (1995) An iron-regulated gene, *magA*, encoding an iron transport protein of *Magnetospirillum* sp. strain AMB-1. J Biol Chem 270:28392–28396. https://doi.org/10.1074/jbc.270.47.28392

Nakazawa H, Arakaki A (2009) Whole genome sequence of *Desulfovibrio magneticus* strain RS-1 revealed common gene clusters in magnetotactic bacteria. Genome Res 19: 1801–1808

Nakazawa H, Arakaki A, Narita-Yamada S et al (2009) Whole genome sequence of *Desulfovibrio magneticus* strain RS-1 revealed common gene clusters in magnetotactic bacteria. Genome Res 19:1801–1808. https://doi.org/10.1101/gr.088906.108

Nguyen HV, Suzuki E, Oestreicher Z et al (2016) A protein-protein interaction in magnetosomes: TPR protein MamA interacts with an Mms6 protein. Biochem Biophys Rep 7:39–44. https://doi.org/10.1016/j.bbrep.2016.05.010

Niggli V, Kaufmann S, Goldmann WH et al (1994) Identification of functional domains in the cytoskeletal protein talin. Eur J Biochem 224:951–957

Nudelman H, Zarivach R (2014) Structure prediction of magnetosome-associated proteins. Front Microbiol 5:9. https://doi.org/10.3389/fmicb.2014.00009

Okamura Y, Takeyama H, Matsunaga T (2000) Two-dimensional analysis of proteins specific to the bacterial magnetic particle membrane from *Magnetospirillum* sp. AMB-1. Appl Biochem Biotechnol 84–86:441–446. https://doi.org/10.1385/ABAB:84-86:1-9:441

Okamura Y, Takeyama H, Matsunaga T (2001) A magnetosome-specific GTPase from the magnetic bacterium *Magnetospirillum magneticum* AMB-1. J Biol Chem 276:48183–48188. https://doi.org/10.1074/jbc.M106408200

Okamura Y, Takeyama H, Sekine T et al (2003) Design and application of a new cryptic-plasmid-based shuttle vector for *Magnetospirillum magneticum*. Appl Environ Microbiol 69:4274–4277. https://doi.org/10.1128/AEM.69.7.4274-4277.2003

Pan W, Xie C, Lv J (2012) Screening for the interacting partners of the proteins MamK & MamJ by two-hybrid genomic DNA library of *Magnetospirillum magneticum* AMB-1. Curr Microbiol 64:515–523. https://doi.org/10.1007/s00284-012-0099-2

Peter BJ, Kent HM, Mills IG et al (2004) BAR domains as sensors of membrane curvature: the amphiphysin BAR structure. Science 303:495–499. https://doi.org/10.1126/science.1092586

Pradel N, Santini C-L, Bernadac A et al (2006) Biogenesis of actin-like bacterial cytoskeletal filaments destined for positioning prokaryotic magnetic organelles. Proc Natl Acad Sci U S A 103:17485–17489. https://doi.org/10.1073/pnas.0603760103

Quinlan A, Murat D, Vali H, Komeili A (2011) The HtrA/DegP family protease MamE is a bifunctional protein with roles in magnetosome protein localization and magnetite biomineralization. Mol Microbiol 80:1075–1087. https://doi.org/10.1111/j.1365-2958.2011.07631.x

Randazzo PA, Yang YC, Rulka C, Kahn RA (1993) Activation of ADP-ribosylation factor by Golgi membranes: evidence for a brefeldin A- and protease-sensitive activating factor on Golgi membranes. J Biol Chem 268:9555–9563

Raschdorf O, Müller FD, Pósfai M et al (2013) The magnetosome proteins MamX, MamZ and MamH are involved in redox control of magnetite biomineralization in *Magnetospirillum gryphiswaldense*. Mol Microbiol. https://doi.org/10.1111/mmi.12317

Raschdorf O, Forstner Y, Kolinko I et al (2016) Genetic and ultrastructural analysis reveals the key players and initial steps of bacterial magnetosome membrane biogenesis. PLoS Genet 12:e1006101. https://doi.org/10.1371/journal.pgen.1006101

Rawlings AE, Bramble JP, Walker R et al (2014) Self-assembled MmsF proteinosomes control magnetite nanoparticle formation *in vitro*. Proc Natl Acad Sci U S A 111:16094–16099. https://doi.org/10.1073/pnas.1409256111

Richter M, Kube M, Bazylinski DA et al (2007) Comparative genome analysis of four magnetotactic bacteria reveals a complex set of group-specific genes implicated in magnetosome biomineralization and function. J Bacteriol 189:4899–4910. https://doi.org/10.1128/JB.00119-07

Scheffel A, Schüler D (2007) The acidic repetitive domain of the *Magnetospirillum gryphiswaldense* MamJ protein displays hypervariability but is not required for magnetosome chain assembly. J Bacteriol 189:6437–6446. https://doi.org/10.1128/JB.00421-07
Scheffel A, Gruska M, Faivre D et al (2006) An acidic protein aligns magnetosomes along a filamentous structure in magnetotactic bacteria. Nature 440:110–114. https://doi.org/10.1038/nature04382
Scheffel A, Gärdes A, Grünberg K (2008) Major magnetosome proteins MamGFDC are not essential for magnetite biomineralization in *Magnetospirillum gryphiswaldense* but regulate the size of magnetosome crystals. J Bacteriol 190(1):377–386
Schübbe S, Würdemann C, Peplies J et al (2006) Transcriptional organization and regulation of magnetosome operons in *Magnetospirillum gryphiswaldense*. Appl Environ Microbiol 72:5757–5765. https://doi.org/10.1128/AEM.00201-06
Schübbe S, Williams T, Xie G (2009) Complete genome sequence of the chemolithoautotrophic marine *magnetotactic coccus* strain MC-1. Appl Environ Microbiol 75:4835–4852
Schultheiss D, Schüler D (2003) Development of a genetic system for *Magnetospirillum gryphiswaldense*. Arch Microbiol 179:89–94. https://doi.org/10.1007/s00203-002-0498-z
Sharma VK, Filip J, Zboril R, Varma RS (2015) Natural inorganic nanoparticles – formation, fate, and toxicity in the environment. Chem Soc Rev 44:8410–8423. https://doi.org/10.1039/C5CS00236B
Shimada A, Niwa H, Tsujita K et al (2007) Curved EFC/F-BAR-domain dimers are joined end to end into a filament for membrane invagination in endocytosis. Cell 129:761–772. https://doi.org/10.1016/j.cell.2007.03.040
Shimizu K, Cha J, Stucky GD, Morse DE (1998) Silicatein alpha: Cathepsin L-like protein in sponge biosilica. Proc Natl Acad Sci 95:6234–6238. https://doi.org/10.1073/pnas.95.11.6234
Siponen MI, Adryanczyk G, Ginet N et al (2012) Magnetochrome: a c-type cytochrome domain specific to magnetotatic bacteria. Biochem Soc Trans 40:1319–1323. https://doi.org/10.1042/BST20120104
Siponen MI, Legrand P, Widdrat M et al (2013) Structural insight into magnetochrome-mediated magnetite biomineralization. Nature 502:681–684. https://doi.org/10.1038/nature12573
Smalley MD, Marinov GK, Bertani LE, DeSalvo G (2015) Genome sequence of *Magnetospirillum magnetotacticum* strain MS-1. Genome Announc 3:e00233–e00215. https://doi.org/10.1128/genomeA.00233-15
Song H-P, Li X-G, Sun J-S et al (2008) Application of a magnetotactic bacterium, *Stenotrophomonas* sp. to the removal of Au(III) from contaminated wastewater with a magnetic separator. Chemosphere 72:616–621. https://doi.org/10.1016/j.chemosphere.2008.02.064
Stenmark H (2009) Rab GTPases as coordinators of vesicle traffic. Nat Rev Mol Cell Biol 10:513–525. https://doi.org/10.1038/nrm2728
Suzuki T, Okamura Y, Calugay RJ et al (2006) Global gene expression analysis of iron-inducible genes in *Magnetospirillum magneticum* AMB-1. J Bacteriol 188:2275–2279
Tanaka M, Okamura Y, Arakaki A et al (2006) Origin of magnetosome membrane: proteomic analysis of magnetosome membrane and comparison with cytoplasmic membrane. Proteomics 6:5234–5247. https://doi.org/10.1002/pmic.200500887
Tanaka M, Nakata Y, Mori T et al (2008) Development of a cell surface display system in a magnetotactic bacterium, *Magnetospirillum magneticum* AMB-1. Appl Environ Microbiol 74:3342–3348. https://doi.org/10.1128/AEM.02276-07
Tanaka M, Arakaki A, Matsunaga T (2010a) Identification and functional characterization of liposome tubulation protein from magnetotactic bacteria. Mol Microbiol 76:480–488
Tanaka M, Arakaki A, Staniland SS, Matsunaga T (2010b) Simultaneously discrete biomineralization of magnetite and tellurium nanocrystals in magnetotactic bacteria. Appl Environ Microbiol 76:5526–5532. https://doi.org/10.1128/AEM.00589-10
Tanaka M, Kawase M, Tanaka T, Matsunaga T (2011a) Gold biorecovery from plating waste by magnetotactic bacterium, *Magnetospirillum magneticum* AMB-1. MRS Proc 1169. https://doi.org/10.1557/PROC-1169-Q03-12

Tanaka M, Mazuyama E, Arakaki A, Matsunaga T (2011b) MMS6 protein regulates crystal morphology during nano-sized magnetite biomineralization *in vivo*. J Biol Chem 286:6386–6392. https://doi.org/10.1074/jbc.M110.183434

Tanaka M, Brown R, Hondow N et al (2012a) Highest levels of Cu, Mn and Co doped into nanomagnetic magnetosomes through optimized biomineralisation. J Mater Chem 22:11919. https://doi.org/10.1039/c2jm31520c

Tanaka M, Critchley K, Matsunaga T et al (2012b) Fabrication of lipid tubules with embedded quantum dots by membrane tubulation protein. Small 8:1590–1595

Tanaka M, Knowles W, Brown R et al (2016) Biomagnetic recovery and bioaccumulation of selenium granules in magnetotactic bacteria. Appl Environ Microbiol 82:3886–3891. https://doi.org/10.1128/AEM.00508-16

Taoka A, Asada R, Wu LF, Fukumori Y (2007) Polymerization of the actin-like protein MamK, which is associated with magnetosomes. J Bacteriol 189:8737–8740. https://doi.org/10.1128/JB.00899-07

Taoka A, Eguchi Y, Mise S et al (2014) A magnetosome-associated cytochrome MamP is critical for magnetite crystal growth during the exponential growth phase. FEMS Microbiol Lett 358:21–29. https://doi.org/10.1111/1574-6968.12541

Uebe R, Junge K, Henn V et al (2011) The cation diffusion facilitator proteins MamB and MamM of *Magnetospirillum gryphiswaldense* have distinct and complex functions, and are involved in magnetite biomineralization and magnetosome membrane assembly. Mol Microbiol 82:818–835. https://doi.org/10.1111/j.1365-2958.2011.07863.x

Ullrich S, Schüler D (2010) Cre-lox-based method for generation of large deletions within the genomic magnetosome island of *Magnetospirillum gryphiswaldense*. Appl Environ Microbiol 76:2439–2444. https://doi.org/10.1128/AEM.02805-09

Wahyudi AT, Takeyama H, Matsunaga T (2001) Isolation of *Magnetospirillum magneticum* AMB-1 mutants defective in bacterial magnetic particle synthesis by transposon mutagenesis. Appl Biochem Biotechnol 91–93:147–154. https://doi.org/10.1385/ABAB:91-93:1-9:147

Wahyudi AT, Takeyama H, Okamura Y et al (2003) Characterization of aldehyde ferredoxin oxidoreductase gene defective mutant in *Magnetospirillum magneticum* AMB-1. Biochem Biophys Res Commun 303:223–229. https://doi.org/10.1016/S0006-291X(03)00303-6

Wang X, Liang L, Song T, Wu L (2009) Magnetosome formation and expression of *mamA*, *mms13*, *mms6* and *magA* in *Magnetospirillum magneticum* AMB-1 exposed to pulsed magnetic field. Curr Microbiol 59:221–26a

Wang X, Wang Q, Zhang W et al (2014) Complete genome sequence of *Magnetospirillum gryphiswaldense* MSR-1. Genome Announc 2:e00171–14–e00171–14. https://doi.org/10.1128/genomeA.00171-14

Weiner S, Addadi L (1997) Design strategies in mineralized biological materials. J Mater Chem 7:689–702. https://doi.org/10.1039/a604512j

Yamagishi A, Tanaka M, Lenders JJM et al (2016) Control of magnetite nanocrystal morphology in magnetotactic bacteria by regulation of *mms7* gene expression. Sci Rep 6:29785. https://doi.org/10.1038/srep29785

Yamamoto D, Taoka A, Uchihashi T et al (2010) Visualization and structural analysis of the bacterial magnetic organelle magnetosome using atomic force microscopy. Proc Natl Acad Sci U S A 107:9382–9387

Yang W, Li R, Peng T et al (2010) MamO and mamE genes are essential for magnetosome crystal biomineralization in *Magnetospirillum gryphiswaldense* MSR-1. Res Microbiol 161:701–705. https://doi.org/10.1016/j.resmic.2010.07.002

Yang J, Li S, Huang X et al (2013) MamX encoded by the *mamXY* operon is involved in control of magnetosome maturation in *Magnetospirillum gryphiswaldense* MSR-1. BMC Microbiol 13:203. https://doi.org/10.1186/1471-2180-13-203

Yoshino T, Shimojo A, Maeda Y, Matsunaga T (2010) Inducible expression of transmembrane proteins on bacterial magnetic particles in *Magnetospirillum magneticum* AMB-1. Appl Environ Microbiol 76:1152–1157. https://doi.org/10.1128/AEM.01755-09

Zeytuni N, Ozyamak E, Ben-harush K et al (2011) Self-recognition mechanism of MamA, a magnetosome-associated TPR-containing protein, promotes complex assembly. Proc Natl Acad Sci U S A 108:480–487. https://doi.org/10.1073/pnas.1103367108/-/DCSupplemental. www.pnas.org/cgi/doi/10.1073/pnas.1103367108
Zeytuni N, Uebe R, Maes M et al (2014) Cation diffusion facilitators transport initiation and regulation is mediated by cation induced conformational changes of the cytoplasmic domain. PLoS One. https://doi.org/10.1371/journal.pone.0092141
Zeytuni N, Cronin S, Lefèvre CT et al (2015) MamA as a model protein for structure-based insight into the evolutionary origins of magnetotactic bacteria. PLoS One 10:e0130394. https://doi.org/10.1371/journal.pone.0130394

Chapter 3
Structural and Proteomic Analyses of Iron Oxide Biomineralization in Chiton Teeth

David Kisailus and Michiko Nemoto

Abstract The denticle caps of chiton teeth exhibit the largest hardness and stiffness among any biological minerals in the world. They consist of a composite structure of highly oriented crystalline nanorods of magnetite surrounded by a veneer of organic. The ultrahard teeth of the chiton are the first reported example of biologically formed magnetite, and its formation mechanisms have been of great interest to researchers for decades. Recently, using modern nanomechanical characterization techniques, it was shown that chiton teeth exhibit the largest hardness and stiffness of any biological mineral. The predicted abrasion resistance of chiton teeth against a blunt contact was reported to be even higher than that of zirconia. Therefore, the elucidation of chiton teeth synthetic processes could help us learn how to develop novel abrasion-resistant materials and environmentally benign processes for the production of iron oxides and other nanostructured materials. In order to understand the chiton teeth formation, proteomic analyses of tooth proteins as well as detailed structural analyses of mature and developing teeth were conducted. Based on the results obtained from these analyses, we discuss the underlying mechanisms of iron oxide biomineralization in chiton teeth.

3.1 Introduction

It has been known that widely distributed magnetite on our planet is lithogenic, originating from igneous rock, which formed at high temperatures and pressures. Other than lithogenic magnetite, pedogenic magnetite that is formed in soil has been recently discovered (Maher and Taylor 1988). However, in 1962, Lowenstam, using

D. Kisailus (✉)
Department of Chemical and Environmental Engineering and Materials Science and Engineering Program, University of California, Riverside, CA, USA
e-mail: david@engr.ucr.edu

M. Nemoto
Graduate School of Environmental and Life Science, Okayama University, Okayama, Japan
e-mail: mnemoto@okayama-u.ac.jp

T. Matsunaga et al. (eds.), *Biological Magnetic Materials and Applications*,
https://doi.org/10.1007/978-981-10-8069-2_3

X-ray diffraction methods, showed for the first time that the denticle caps of chiton teeth consisted of magnetite (Lowenstam 1962). This was the first report of biologically formed magnetite and previously believed to be formed only under elevated temperatures and pressures. After this initial finding, a plethora of biologically mineralized magnetite has been found from other organisms including bacteria, homing pigeons, honeybees, and salmon (Blakemore 1975; Gould et al. 1978; Kirschvink et al. 1985; Walcott et al. 1979). In fact, although the origin of pedogenic magnetite is still under discussion, biotic routes to magnetite have been suggested as a possible route (Fassbinder et al. 1990). In order to mineralize the magnetite under physiological conditions, it was suggested that organisms might use a highly regulated system to control oxidation/reduction of iron oxides.

In addition to offering a low temperature, environmentally benign biological process to magnetite, the controlled synthesis of this ultrahard material has afforded another advantage, superior abrasion resistance. This notable feature was revealed in the teeth of chiton, an algae-grazing mollusk. For many years, Chitons have been known to be the cause of erosion of rocky surfaces in the littoral zones. The great erosional power of chiton can be found from deep gauges on these rocks (primarily limestone, 3–4 on Moh's hardness scale). In order to access algae growing on (epilithic) and within (endolithic) rock, the teeth of the chiton must be able to abrade the rock without becoming worn prematurely. Lowenstam was the first to show that the hardness of chiton teeth was close to 6 on Moh's hardness scale (Lowenstam 1962). Recently, using modern nanomechanical characterization techniques, Weaver et al. showed that chiton teeth exhibit the largest hardness and stiffness among any biological minerals. The predicted abrasion resistance of chiton teeth against a blunt contact was reported to be even higher than that of zirconia, which has been used in abrasion-resistant coatings and other load-bearing applications (Weaver et al. 2010). Recent analysis using microscopic and spectroscopic techniques combined with finite element simulations revealed the relationships between structure and mechanical properties of chiton teeth (Grunenfelder et al. 2014). In this work, the magnetite in shell of the teeth was shown to consist of nanorods oriented along the long axis of the tooth, providing superior abrasion resistance and strength under tensile loading on the leading edge as well as under compression loading on the trailing edge. This was due to the highly controlled nature of the synthesis process.

In this chapter, we will concentrate on aspects relevant to the formation mechanisms of chiton teeth. Unanswered questions and future research directions of chiton teeth formation are highlighted through the review of previous literature. The analysis of chiton teeth could lead to the elucidation of magnetite biomineralization mechanisms shared by all magnetite-forming organisms. Furthermore, it could pave the way for developing novel environmentally benign processes for the production of iron oxides and other nanostructured materials.

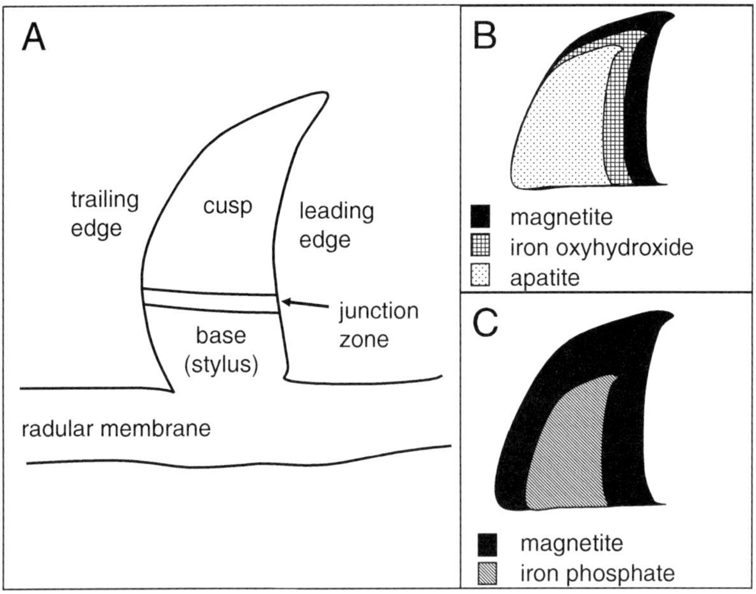

Fig. 3.1 Schematic illustration of the radular teeth of chiton. (**a**) Basic structure of a radular tooth. Composition of the mineralized cusp in (**b**) *Acanthopleura* and (**c**) *C. stelleri*

3.2 Structure and Composition of Fully Mineralized Chiton Teeth

Several chiton species that inhabit both cold and warm waters have been used to study the formation of magnetite within their teeth. Among them, *Cryptochiton stelleri*, which inhabits the North Pacific Ocean, is the world largest species, reaching a maximum length of more than 30 cm. Because of its size, this species is an ideal research target to study the formation mechanisms of magnetite within chiton teeth. *C. stelleri* has more than 70 primary rows of teeth (termed major lateral teeth) that are attached to a ribbon-like structure called a radular membrane. Each of the 70 rows consists of one pair of teeth; the anterior teeth on the membrane are fully mineralized and subsequently have reduced mineral content moving toward the posterior. Each tooth is composed of a mineralized cusp and base (also referred to as stylus) supporting the mineralized cusp (Fig. 3.1a). The morphology of teeth is different among species. *C. stelleri* and *Plaxiphora albida* have tricuspid teeth (Fig. 3.2b), while other chiton species including *Acanthopleura* has unicuspid teeth. Adjacent the major lateral teeth, marginal teeth are positioned in ridges along the side of the stylus to provide lateral support during rasping. The entire structure, including the mineralized major lateral teeth, marginal teeth, and the radular membrane, is called the radula (Fig. 3.2b).

Chiton use only several rows (~10) of teeth in the most anterior region for grazing on rock for algae. When these teeth are worn down, they break off, and newly

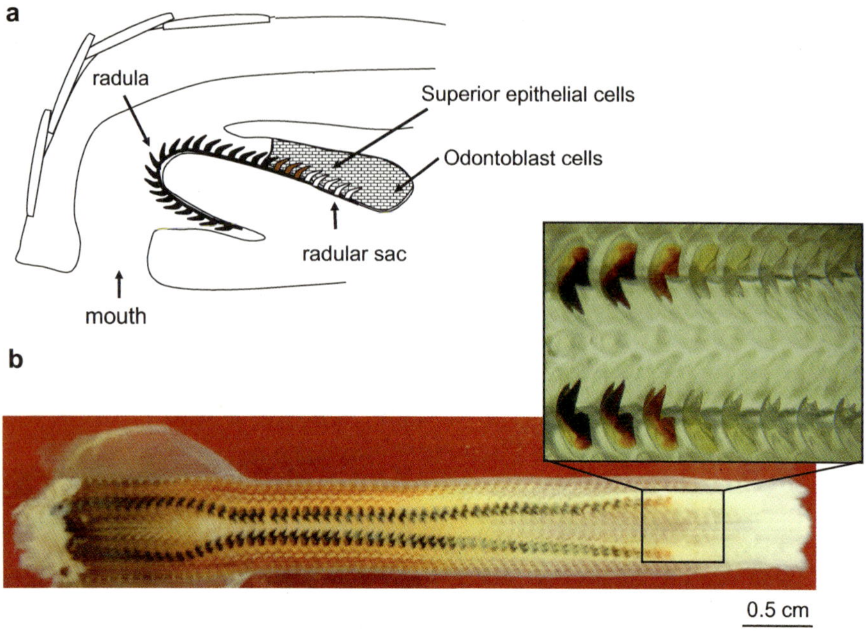

Fig. 3.2 (**a**) Diagram of the mouth part of a chiton; (**b**) optical micrograph of the radula of *C. stelleri*

formed teeth move forward to replace them (Fig. 3.2a). Radular teeth are shaped inside an organ called the radular sac. During tooth formation, the tooth matrix, which is composed of α-chitin and protein, is initially secreted from odontoblast cells associated with the posterior end of the sac. Subsequently, mineralization occurs within the tooth matrix through several steps to form mature teeth. The formation rate of each new row of teeth varies from 0.4 to 0.8 rows per day (Nesson 1968; Shaw et al. 2008) and is dependent on their feeding habits. Since the teeth in different mineralization stages exist at the same time on one radula, it is an attractive research model to study dynamic biomineralization processes.

The pioneering work of Lowenstam showed that an individual mineralized tooth is a composite material, consisting of several mineral phases including magnetite, lepidocrocite, and apatite (Lowenstam 1967). Mineral distribution in each tooth has been analyzed by a number of methods including transmission electron microscopy (TEM), energy dispersive spectrometry (EDS), and laser Raman microscopy (Lee et al. 1998, 2000; Saunders et al. 2009). Analyses of the teeth from *Acanthopleura echinata* and *Acanthopleura hirtosa* have been revealed an outermost shell of magnetite, followed by thin layers of iron oxyhydroxide phases (lepidocrocite or lepidocrocite plus goethite) and a core consisting of apatite (Fig. 3.1b). The outermost magnetite cap is common in all chiton species, while the composition of the core region varies among species. A number of chiton species, including *Acanthopleura*, have apatite at the core of their teeth (Evans et al. 1992; Lee et al. 2000; Lowenstam

1967; Lowenstam and Weiner 1985). The detailed analysis using Fourier transform infrared spectroscopy (FTIR), Raman spectroscopy, and X-ray diffraction revealed the apatite phases consisted of carbonated hydroxyapatite, dahllite, or carbonated fluorapatite (francolite) (Evans et al. 1992; Lee et al. 2000; Lowenstam and Weiner 1985). *C. stelleri* have weakly crystalline-hydrated iron phosphate (phosphosiderite) in its core region (Fig. 3.1c) (Weaver et al. 2010). Amorphous hydrous ferric phosphate was also detected in the teeth of *Mopalia muscosa* (Nesson and Lowenstam 1985). *Plaxiphora albida* was found to have limonite, which is a mixture of hydrated lepidocrocite and goethite, in its core region (Lee et al. 2003). Mössbauer spectroscopic analysis of *A. hirtosa* further revealed the presence of maghemite, which is the oxidized form of magnetite in the mature teeth (Mizota and Maeda 1986; Pierre et al. 1992). In *A. hirtosa*, maghemite was estimated as ~10% of total Fe in the fully mineralized teeth of the radulae (Pierre et al. 1992). It was suggested that part of magnetite was oxidized to maghemite in radular teeth.

Elemental composition of radular teeth has been analyzed by a number of methods including proton-induced X-ray emission (PIXE) (Kim et al. 1986b). One such analysis showed that the concentrations of Fe, Ca, P, and F increased as mineralization proceeded toward the anterior end of the radula of *A. hirtosa*. Additionally, trace amounts of Zn also increased through the course of mineralization. But the role of Zn for tooth mineralization has yet to be determined. An inductively coupled plasma mass spectrometry (ICP-MS) (Okoshi and Ishii 1996) or inductively coupled plasma atomic emission spectroscopy (ICP-AES) (Shaw et al. 2008) was also used to analyze the elemental composition of radular teeth. Beyond the dominant elements found in the mineral of teeth, Fe, Ca, O, and P, relatively high concentrations of Al, Mg, Na, K, Sr, and Zn were detected (Macey and Brooker 1996; Weaver et al. 2010). The concentrations of minerals among chiton species vary. Among those elements, the content of Mg was decreased after protease treatment (Okoshi and Ishii 1996). This indicated that Mg could be a component of protein structures within the teeth.

Beyond the elemental and phase differences, the volume of mineral deposits and their architecture are different between leading and trailing edges in different species. More minerals are deposited on leading edge than on trailing edge of the tooth cusp in *A. hirtosa* (Kim et al. 1989). Wang et al. revealed a structural gradient in the teeth of chiton *C. stelleri* (Wang et al. 2013). The fracture surface analysis of teeth in *C. stelleri* showed a regionally segregated complex structure consists of (a) thin zone of condensed magnetite nanoparticles on the surface of leading edge, (b) bundles of magnetite rods oriented parallel to the leading edge, (c) those oriented parallel to the trailing edge, and (d) the rod's gradual bending around the core region (Fig. 3.3) (Wang et al. 2013). The diameter of magnetite rods differs in leading edge (162 ± 22 nm) and trailing edge (194 ± 30 nm), affording abrasion resistance and reducing catastrophic failure in the teeth. How these structural gradients are formed inside the tooth was revealed by a detailed TEM analysis of developing teeth, which is explained in the next section.

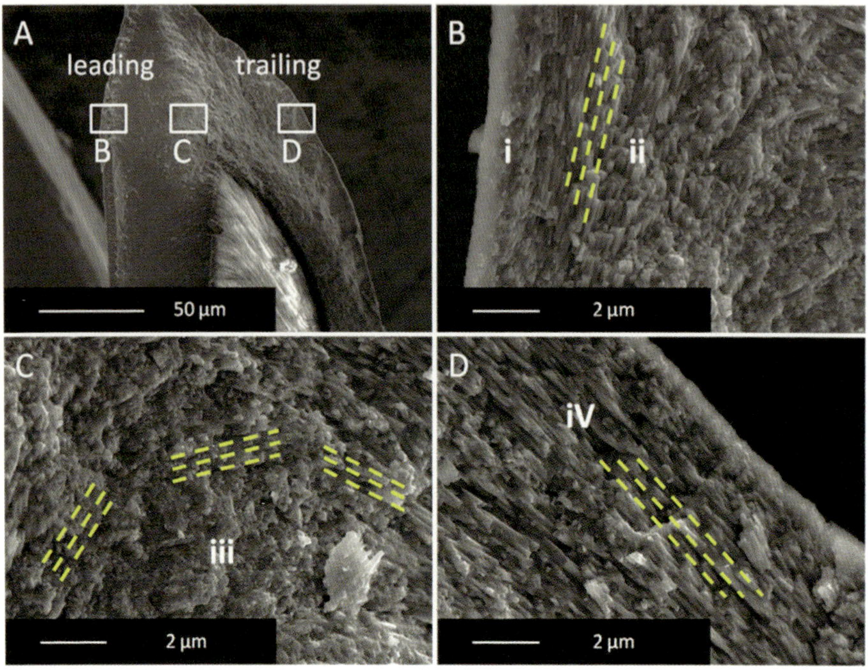

Fig. 3.3 (**a–d**) Scanning electron micrographs of a longitudinal fracture surface of a mineralized radular tooth in *C. stelleri*. Regional structural heterogeneity includes (i) the granular magnetite veneer, the magnetite rods oriented parallel to the (ii) leading and (iv) trailing tooth edges and (iii) a gradual bending at transitional zone between them. The *dashed lines* illustrate the prevalent orientation of the rods in regions (ii), (iii), and (iv) (Reprinted with permission from Wang et al. 2013. Copyright 2013 WILEY-VCH Verlag GmbH & Co. KGaA, Weinheim)

Finally, the volume of mineral on the leading and trailing edges of these teeth, as well as their varied structural components, is likely a function of the specific feeding habits of each species.

3.3 Iron Oxide Mineralization in Chiton Teeth

3.3.1 Phase Transformation of Iron Oxides

As described in Sect. 3.2, chiton possesses all stages of tooth development in one radula. Phase analysis within each region of radular teeth was determined using a combination of traditional powder and synchrotron X-ray diffraction, TEM, and Raman spectroscopy. For *C. stelleri*, the development stages are as follows.

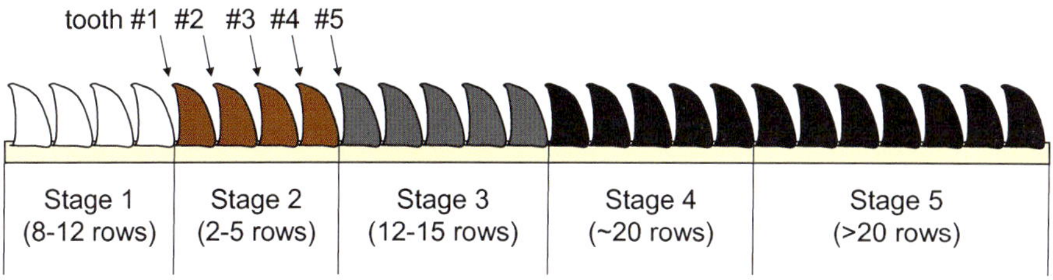

Fig. 3.4 Schematic diagram of the radula in the chiton *C. stelleri* showing every stage of tooth development is contained on a single radula

Stage 1: The first 8–12 rows of teeth, newly secreted from odontoblast cells, are transparent and are composed of α-chitin and proteins and do not contain mineral.
Stage 2: The next 2–5 rows of teeth have a reddish-brown color due to the initial deposition of amorphous iron oxide or ferrihydrite inside the teeth.
Stage 3: The following 12–15 rows of teeth have black-colored cusps. Here, transformation of ferrihydrite to magnetite begins to take place. Ferrihydrite is still detected in this stage. The core region remains void of any mineral.
Stage 4: The subsequent 20 rows of teeth have nearly fully mineralized magnetite cusps, and the core region begins to be filled with iron phosphate.
Stage 5: Finally, the last 20 or more rows of teeth become fully mineralized in both shell and core.

As previously mentioned, in order to identify the mineral phases, teeth in the early developing stage (stage 2 teeth in *C. stelleri*) have been analyzed by X-ray diffraction as well as by electron diffraction in TEM. Results revealed the presence of mineral deposits which had a diameter of ~200 Å. Although X-ray diffraction data was performed, the mineral phase of the deposits was not determined at that time (Towe and Lowenstam 1967). Cornell and Schwertmann used X-ray diffraction to confirm the mineral phase is in good agreement with both synthetic and natural six-line ferrihydrite exhibiting the major peaks corresponding to the (110), (112), and (300) lattice planes (Cornell and Schwertmann 2003). Ferrihydrite has also been detected from developing teeth in other chiton species. TEM analysis of partially mineralized teeth in *A. hirtosa* revealed the deposition of small particle aggregates between 15 and 50 nm on an organic structure (Kim et al. 1989). Electron diffraction analyses indicated the presence of ferrihydrite (Fh).

In order to obtain a more detailed analysis of the phase transformation during the tooth development, the sequential early mineralization stages of teeth were analyzed by synchrotron X-ray techniques (Wang et al. 2013). In their analysis, the first tooth row of stage 2 that showed the earliest visible signs of mineralization was labeled as tooth #1. The following four rows of teeth are numbered sequentially as seen in Fig. 3.4. The results showed that multiple mineral phases coexisted in the early stages of partially mineralized teeth (Fig. 3.5a). Furthermore, ferrihydrite appeared only in the first four rows of mineralized teeth. This suggested that the

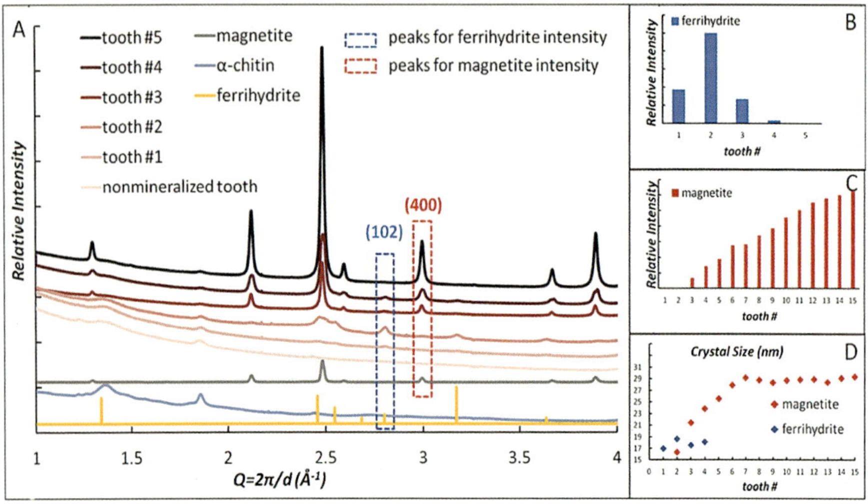

Fig. 3.5 Phase transformations during the early stages of radula mineralization in *C. stelleri.* (**a**) Representative diffracted synchrotron X-ray intensity vs Q (Q = 2π/d) plots from azimuthally integrated patterns in X-ray diffraction frames. Spectra show powder diffraction patterns from a nonmineralized tooth; teeth #1, #2, #3, and #5; magnetite and α-chitin standards; and a ferrihydrite literature reference (Jansen et al. 2002). Diffraction peaks Q = 2.802 $Å^{-1}$ (specific to the 102 ferrihydrite reflection) and Q = 2.991 $Å^{-1}$ (specific to the 400 magnetite reflection) were fit to obtain the tooth-dependent integrated reflection areas and the full-width at half maximum. The intensity of (**b**) ferrihydrite and (**c**) magnetite is represented by the integrated peak areas. (**d**) The change of the average crystal size of magnetite and ferrihydrite was calculated from the full width at half maximum of representative reflections via the Scherrer equation (Reprinted with permission from Wang et al. 2013. Copyright 2013 WILEY-VCH Verlag GmbH & Co. KGaA, Weinheim)

phase transformation from ferrihydrite to magnetite occurred relatively quickly, within four rows of teeth (Wang et al. 2013) (Fig. 3.5b). X-ray diffraction patterns from a synchrotron beam line further enabled the quantification of crystal size variation of each mineral phase in the partially mineralized teeth. The results showed that there were no obvious changes in ferrihydrite crystal size (17–18.7 nm) during the tooth mineralization, while the magnetite crystals grew from ~16.4 to ~29.2 nm (Fig. 3.5d). Phase transformation from ferrihydrite to magnetite in the early stages of mineralization was also observed by Mössbauer spectra (Mizota and Maeda 1986), also indicating the size development of magnetite particles occurred along with tooth development.

Wang et al. conducted synchrotron micro X-ray fluorescence (μXRF) analysis and micro X-ray absorption near-edge structure (μXANES) analysis to construct spatially resolved maps of ferric distribution in early stages of partially mineralized teeth. According to their results, ferrihydrite first deposits within the leading edge of the teeth as well as at the interface between the tooth and base. Subsequently, ferrihydrite localized at the tip of the leading edge starts to transform to magnetite (Fig. 3.6). Initial iron deposition at the interface between tooth and base (junction zone) was also reported in *A. hirtosa* (Shaw et al. 2009a).

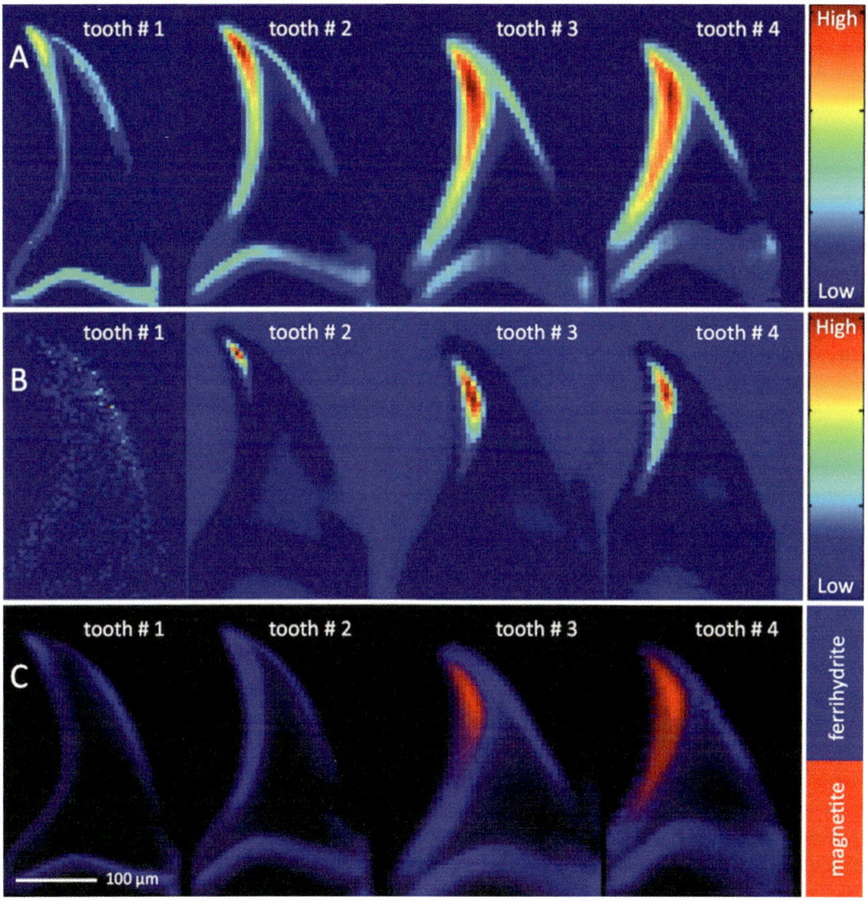

Fig. 3.6 Micro X-ray fluorescence (µXRF) analysis of longitudinal thin sections of radular teeth #1–4 from *C. stelleri* shows the local iron oxide transformation. Both the intensity of (**a**) total iron and (**b**) ferrous iron increase across the leading edge with increasing tooth maturation. (**c**) Micro X-ray absorption near-edge structure (µXANES) analysis reveals an increasing abundance of magnetite across the shell region along the leading edge of radular teeth (Reprinted with permission from Wang et al. 2013. Copyright 2013 WILEY-VCH Verlag GmbH & Co. KGaA, Weinheim)

Recently, iron-organic complexes formed prior to Fh formation were detected in stage 1 teeth by spectroscopic analysis. Gordon et al. used X-ray absorption analysis and electron paramagnetic resonance spectroscopic analysis to detect a mononuclear high-spin Fe ^{III} iron complex in stage 1 teeth of chiton *K. tunicate* (Gordon et al. 2014). In that report, they also showed that polyaspartic acid or citrate in a $FeCl_2$ solution induced the formation of ferrihydrite and high-spin Fe ^{III} iron complexes. Based on these results, the acidic ligands were suggested to be involved in the formation of Fe ^{III} iron complex before Fh formation.

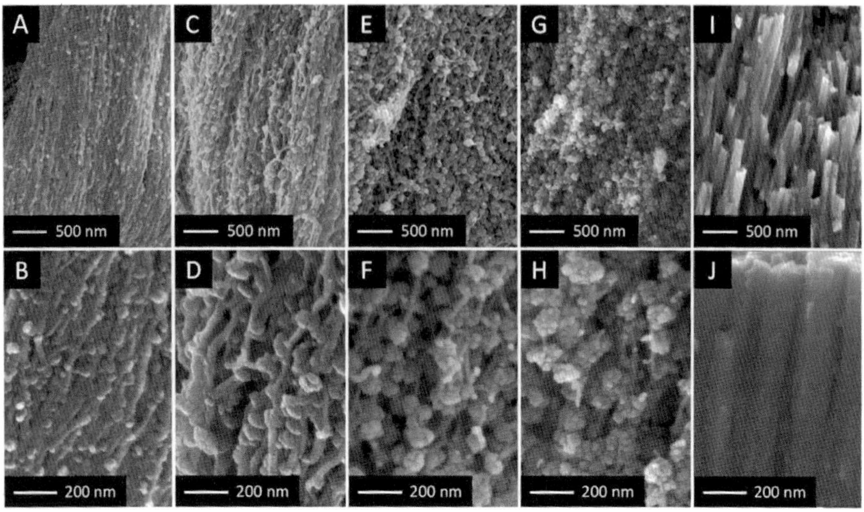

Fig. 3.7 Scanning electron micrographs of fracture surfaces from the tips of (**a**, **b**) tooth #1, (**c**, **d**) tooth #2, (**e**, **f**) tooth #3, (**g**, **h**) tooth #4 and (**i**, **j**) a fully mineralized tooth in *C. stelleri* (Reprinted with permission from Wang et al. 2013. Copyright 2013 WILEY-VCH Verlag GmbH & Co. KGaA, Weinheim)

3.3.2 Ultrastructural Development

Fracture surface analysis of the tips of teeth #1 through #4 (Fig. 3.4) showed that the crystal aggregates first deposited on alpha-chitin fibers continuously grow in size as teeth maturation proceeds (Fig. 3.7). In tooth #1, 41 ± 6 nm crystal aggregates of Fh were first deposited along the fiber. The size of these aggregates gradually increases to 107 ± 18 nm. With the growing size of crystal aggregates, the diameter of organic fibers also was observed to decrease from 37 ± 2 to 20 ± 3 nm.

TEM analysis revealed that size and density of mineral deposits in tooth #3 differ between the tip, middle, and bottom of the tooth leading edge (Fig. 3.8b–d). The average mineral deposit size was reported to be 175.0 ± 24 nm at the tip, 70.7 ± 14 nm at the middle, and 40.5 ± 8 nm at the base of the leading edge. The number density of mineral deposit was $6.4/\mu m^2$ at the tip, $66.9/\mu m^2$ at the middle, and $106.2/\mu m^2$ at the base. Furthermore, at the tip of tooth #3, the mineral deposited on the leading edge was smaller and denser compared to those deposited on the trailing edge (Fig. 3.8b, f). These regional differences were suggested to be controlled by the organic framework, alpha-chitin (the detailed description about the chitin in chiton teeth is in Sect. 3.3.3.1).

High-resolution TEM observations of the tip of tooth #5 revealed magnetite nanocrystal aggregates composed of ~16 to 19 nm multiple nanocrystals surrounded by larger magnetite crystals, which were 56–128 nm (Fig. 3.8h). Since the size range of magnetite nanocrystals was similar to that of ferrihydrite initially deposited on teeth #1 to #4 (Fig. 3.5d), a solid-state phase transformation from ferrihydrite to

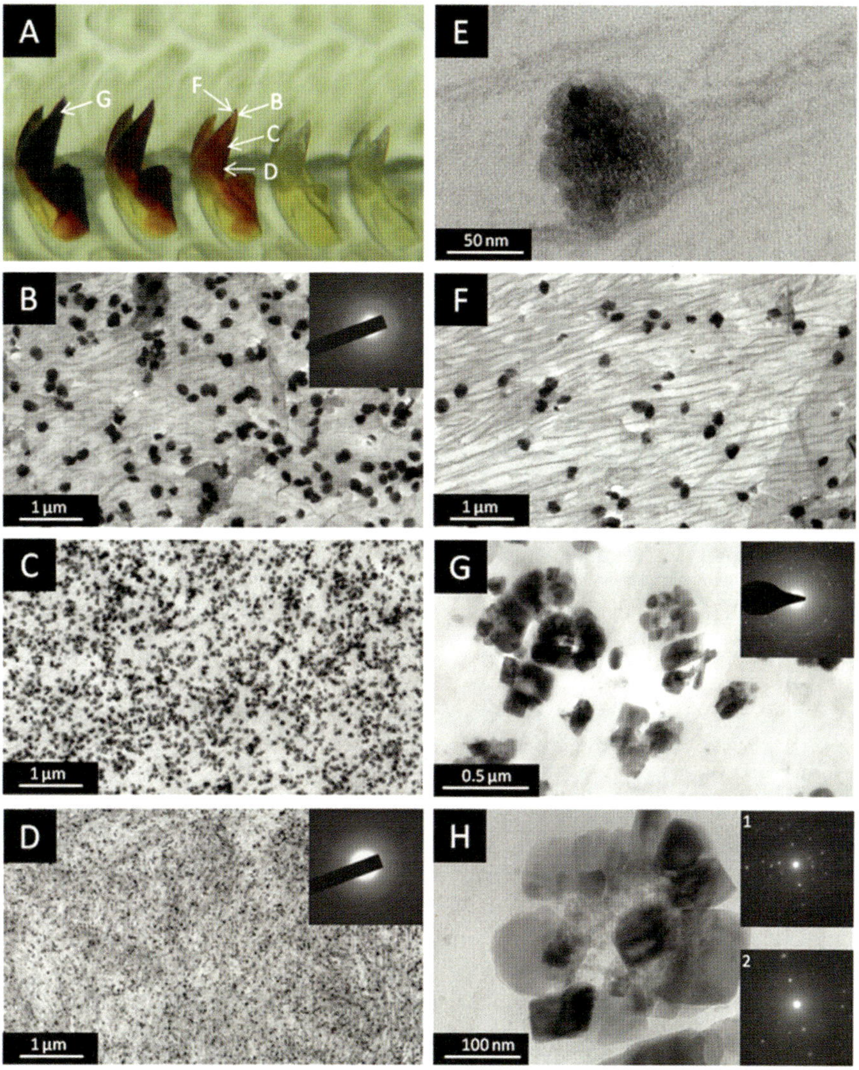

Fig. 3.8 Transmission electron microscopy (TEM) analyses of phase transformations during early radular tooth mineralization (i.e., teeth #3 and #5) in *C. stelleri.* (**a**) An optical micrograph illustrates the locations of TEM sections. TEM from (**b**) the tip, (**c**) the middle, and (**d**) the base of the leading edge of tooth #3. Selected area electron diffraction (SAED) shows the presence of ferrihydrite and magnetite at the tooth tip, while the nanocrystalline aggregates at the base only contain ferrihydrite. (**e**) Deposited mineral at the trailing edge of the tip of the tooth is physically contacted with the fibrillar organic matrix and consists of aggregated nanocrystallites with 16–19 nm in diameter. (**f**) TEM from the tip of the trailing edge of tooth #3. (**g**, **h**) Magnetite crystal aggregates from the tip of tooth #5 consist of multiple nanocrystallites with 16–19 nm in size, surrounded by larger single crystals of magnetite. Convergent beam electron diffraction (CBED) illustrates that both the (H, *inset1*) 16–19 nm crystals in the cores of the crystal aggregates and the (H, *inset2*) larger domains along their perimeter are single magnetite crystals (Reprinted with permission from Wang et al. 2013. Copyright 2013 WILEY-VCH Verlag GmbH & Co. KGaA, Weinheim)

magnetite was suggested. The solid-state phase transformation has been suggested to be induced by the adsorption of Fe^{2+} to ferrihydrite (Tronc et al. 1992).

3.3.3 Organic Matrix in Chiton Tooth

Organic matrices such as polysaccharides or proteins have been known to control biomineralization via templating heterogeneous nucleation events. For chiton, it is likely that the organic matrix composed of α-chitin and proteins controls nucleation and growth of mineral during teeth formation. Here, we review previous reports and discuss the possible role of these organic matrices on mineral formation in chiton teeth.

3.3.3.1 Chitin Fibers

From TEM analysis of stained, nonmineralized teeth of *C. stelleri*, Towe et al. found rope-like fibrous materials within the teeth. It was also determined that those organic fibers were embedded within magnetite crystals in fully mineralized teeth (Towe and Lowenstam 1967). Fracture surface analysis of fully mineralized teeth revealed a polygonal organic framework enclosed in magnetite in *C. stelleri*, *Henleya henleyi*, and *Chiton tuberculatus* (Kirschvink and Lowenstam 1979; Towe and Lowenstam 1967). The organic fibers in the teeth were subsequently revealed to be α-chitin, as determined by X-ray diffraction analysis or Raman spectroscopic analysis (Evans et al. 1990; Lowenstam and Weiner 1985; Nesson and Lowenstam 1985; Weaver et al. 2010).

α-Chitin, which is a polymer of N-acetyl-β-d-glucosamine, is the most abundant polymorphic form of chitin. It is found within insect skins, crustacean shells, and fungal or yeast cell walls. The templating activity of chitin in biomineralization has been previously identified (Matsumura et al. 2015).

Ultrastructural analysis using SEM and TEM revealed the close association between chitin and mineral in chiton teeth (Wang et al. 2013). High-resolution TEM revealed that each aggregate of nanocrystals forms directly on the chitin fibers within partially mineralized teeth (Fig. 3.8e). At the tip of tooth #3, the spacing between chitin fibers at the trailing edge was 116 ± 35 nm, while those at the leading edge was 106 ± 24 nm. At the tip of the trailing edge, the size of crystal aggregates was 211.0 ± 29 nm and had a crystal aggregate density of $2.6/\mu m^2$, while those at the leading edge were 175.0 ± 24 nm and $6.4/\mu m^2$, respectively. Based on these observations, it is likely that densely packed chitin fibers in the leading edge provide more mineral nucleating sites resulting in the deposition of larger number of smaller crystal aggregates, while sparsely spaced chitin fibers in the trailing edge provide fewer mineral nucleating sites resulting in the deposition of fewer, yet large crystal aggregates. The difference in the organization of chitin fibers in the trailing and leading edges was also observed in the chiton *Plaxiphora albida* (Macey et al. 1996) and *A. hirtosa* (Evans et al. 1990, 1994). Evans et al. showed that the chitin fibers in the anterior region of the cusp consisted of hollow tubules and appeared to be interconnected by small-diameter organic fibers (Evans et al. 1990). These results, together with the observations obtained from fracture surface analysis of mineralized teeth (Sect. 3.2), suggest that the spacing between chitin fibers determines the final magnetite rod diameter and thus the function in mineralized teeth.

3.3.3.2 Protein

The close observation of developing chiton teeth allows us to assume that there may be several proteins controlling iron oxide mineralization in chiton teeth. These proteins may include the proteins that (1) induce heterogeneous nucleation of ferrihydrite on chitin fibers, (2) control the reduction or oxidation environment for ferrihydrite and/or magnetite formation, and (3) control the transport of iron and other elements.

Histological staining of a nonmineralized teeth section has indicated the presence of the proteins inside the teeth (Evans et al. 1990, 1991). Further analysis revealed that protein content of major lateral teeth in *A. hirtosa* was 10.2% (dry weight). Amino acid analysis suggested that the proteins in radular teeth were rich in acidic amino acids and glycine (Evans et al. 1991). However, since these data were obtained from whole radulae, including the mineralized cusp, radular membrane, and base, all of these proteins were not necessarily involved in iron oxide mineralization.

Gordon and Joester analyzed chiton tooth with pulsed-laser atom-probe tomography, which could reconstruct three-dimensional tomograms of materials with atomic resolution (Gordon and Joester 2011). The results showed that Na^+ and Mg^{2+} co-localize with organic fibers. Since the chitin that constitutes the organic fibers in chiton tooth is not negatively charged and might have low affinity to these cations, they suggested the presence of acidic proteins that could bind with the cations. Together with previous reports, these results indirectly support the presence of acidic proteins in the teeth.

Proteomic Analysis

In order to identify the proteins or peptides that specifically exist in the cusp region controlling magnetite mineralization, proteomic analysis of tooth proteins in *C. stelleri* was conducted (Nemoto et al. 2012). To compare the protein profile in the mineralized cusp and those in the other parts of radular teeth (base + radular membrane), mineralized cusps were physically separated from each base and radular membrane. Subsequently, proteins were extracted from each part of radular teeth using boiling SDS. The protein contents of mineralized tooth cusp and membrane plus radular membrane were 0.19 ± 0.01% and 2.73 ± 0.12% by dry weight, respectively. Not surprisingly, these fractions were significantly lower than those found by

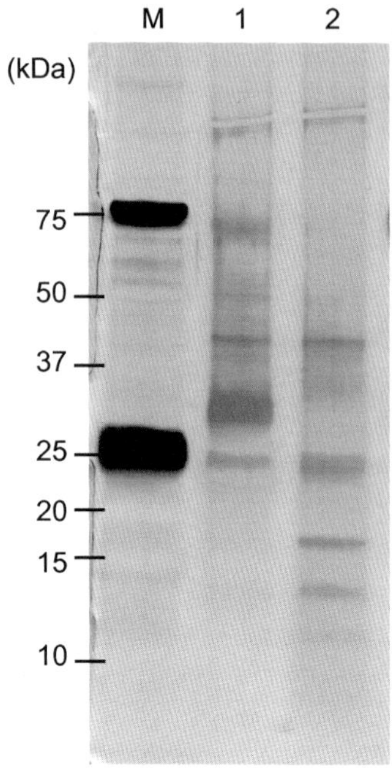

Fig. 3.9 One-dimensional electropherograms of proteins extracted from the radular teeth of the chiton *C. stelleri.* M, Marker; Lane 1, proteins from membrane plus base of mineralized cusp; Lane 2, proteins from mineralized cusp (Nemoto et al. 2012. Copyright 2012 WILEY-VCH Verlag GmbH & Co. KGaA, Weinheim)

Evans et al. (1991). SDS-PAGE results showed several dominant protein bands specific to the mineralized cusp fraction (Fig. 3.9). The proteins extracted from each of the bands were analyzed with nano-LC-MS. As a result, six proteins including the proteins that might be important for iron oxide mineralization were identified as the mineralized cusp-specific proteins (Table 3.1).

A protein showing similarities with dienelactone hydrolase from *A. hirtosa* was identified. Hydrolase is known for its importance for silica biomineralization. The hydrolase isolated from sponge spicule catalyzes the condensation of silicon oxide during sponge spicule formation (Cha et al. 1999). In iron oxide formation, ferrihydrite, which has been detected from partially mineralized teeth as an intermediate mineral phase, is formed through hydrolysis of iron (III) hydroxide precursor. In vitro, the hydrolysis reaction is induced by an addition of a base, by heating or by dilution (Cornell and Schwertmann 2003).

The myoglobin was identified as one of the major proteins in the mineralized cusp fractions. Myoglobin that could reversibly bind with oxygen is known to control the intracellular and muscle oxygen levels. In iron oxide formation, the oxidation rate affects the mineral phase of the final product (Cornell and Schwertmann 2003). Thus the myoglobin identified from the mineralized cusp may be involved in oxygen level control during iron oxide formation.

Other proteins include an arginine kinase that is known to catalyze the transfer of a phosphoryl group of ATP to L-arginine and cyclophilin A that has peptidyl-prolyl cis-trans isomerase activity (Safavi-Hemami et al. 2010; Wang and Heitman 2005). The indirect function in the iron oxide biomineralization of these proteins was

Table 3.1 List of the mineralized cusp-specific proteins from radular teeth in *C. stelleri*

Annotation	Function	Organisms	Accession no.
Full=Globin-1; AltName: Full=Myoglobin I	Oxygen binding	*Liolophura japonica*	Q7M416
Arginine kinase	Arginine kinase activity	*Cellana grata*	BAB41096
		Sepioteuthis lessoniana	BAA95610
		Liolophura japonica	O15990
Dienelactone hydrolase family protein	Hydrolase activity	*Acanthopleura hirtosa*	GO924919
Cyclophilin A	Peptidyl-prolyl cis-trans isomerase activity	*Mythimna separata*	ADI58372
		Acanthopleura hirtosa	GO924939
Glyceraldehyde phosphate dehydrogenase (GAPDH)	Glycolysis and glyconeogenesis	*Sporothrix schenckii*	ACY38586
		Gremmeniella laricina	AAF97478
Eukaryotic translation initiation factor 4A	Translation initiation factor activity	*Drosophila melanogaster*	CAA48790

Table 3.2 List of the mineralized cusp-specific peptides from radular teeth in *C. stelleri*

De novo peptide sequences	m/z	z
SPGNLLVSTR	522.33	2
VLDSLQLTATK	594.89	2
SLGSFYFGAGSAK	646.36	2
ALSTALHEASGLR	663.41	2
LYGWLSTVMLR	669.91	2
LYGWLSTVYPR	677.89	2
LPEVSDLQFDMEK	775.92	2
FLVNVGFANLTPGLK	795.51	2
HLWSFSEDDDSAEK	833.38	2

expected (such as phosphorylation or assisting the assembly/folding of biomineralization-related proteins).

The peptides that did not show homology to any known proteins were further listed and analyzed (Table 3.2). One of the peptides (HLWSFSEDDDSAEK) has a characteristic sequence with its high acidic amino acid content. The protein Mms6 isolated from magnetotactic bacteria has been known to regulate the magnetite crystal structure (Amemiya et al. 2007; Arakaki et al., 2003, 2010; Tanaka et al. 2011). The previous studies suggested that the domain containing the cluster of negatively charged acidic amino acids in Mms6 protein is responsible for iron binding (Arakaki et al. 2003). Thus, the acidic peptides identified from chiton teeth could interact with iron for iron oxide deposition in the tooth cusp. Another peptide identified from the mineralized cusp shares a homology to the peroxiredoxin-6-like protein. The peroxiredoxin-6 protein is known to be involved in redox regulation in cells (Choi et al. 1998). Therefore, the peroxiredoxin-6-like protein in the mineralized cusp could be involved in the reduction of iron oxide species.

The proteomic analysis revealed, for the first time, the protein profile of the iron oxide-based chiton teeth. This data provided the basis for the molecular biological analysis of chiton teeth formation.

3.3.4 Role of Epithelial Cells

During the teeth formation inside the radular sac, entire teeth are surrounded by superior epithelial cells that might be responsible for the transport of iron and other biomolecules required for tooth mineralization (Fig. 3.2a) (Kim et al. 1989; Nesson 1968). Ultrastructural analysis of teeth and superior epithelial cells in each of the mineralization stages was conducted on *A. hirtosa* (Kim et al. 1989). In that report, no Fe-containing materials were detected by Perls' reaction in the superior epithelial cells of the nonmineralized teeth. In the superior epithelial cells, just before the onset of tooth mineralization, large granules containing Fe started to appear. Those Fe-containing granules were distributed uniformly in the epithelial cells attached to the tooth surfaces of both leading and trailing edges. Besides Fe-containing granules, ferritin cores were also observed near the cell membrane of epithelial cells. Characterization of the ferritin isolated from chiton is described in the next section in detail. Furthermore, a number of mitochondria and microvilli attached to the tooth surface were observed. These mitochondria could provide the reductive energy to drive iron release from ferritin cores, while microvilli could be involved in iron transport to the mineralizing surface of tooth cusp (Kim et al. 1989). Similar observations were reported by Shaw et al., who used microwave-assisted sample preparation methods for TEM analysis of partially mineralized teeth in *A. hirtosa* (Shaw et al. 2009b). Comprehensive analysis with a number of partially mineralized teeth enabled the observation of the increased number and length of microvilli attached to the tooth cusp and an increase in the number of granules in superior epithelial cells around the tooth through the course of tooth mineralization. Furthermore, ferritin-like Fe-containing particles were observed in the microvilli attached to the surface of partially mineralized teeth cusps (Shaw et al. 2009b). However, this result contradicts the report from Nesson et al. (Nesson and Lowenstam 1985), which showed no electron-dense material within the microvilli. Further analysis is needed to assess the function of the microvilli for iron transport.

3.3.5 Iron Delivery Pathway

The iron containing protein shell, ferritin, was detected and first isolated from epithelial cells of the radula in *C. stelleri* (Towe et al. 1963). TEM analysis revealed the membrane-enclosed granules composed of paracrystalline or crystalline ferritins appearing in the superior epithelial cells (Lowenstam 1967). After its first appearance in the epithelial cells located one to three rows posterior to the tooth in which first iron oxide deposition occurred, the number of granules increased greatly as tooth mineralization proceeded (Nesson and Lowenstam 1985). This indicates that those ferritins containing granules might be involved in iron transport for tooth mineralization. The size of ferritins was ~5.5–6 nm in diameter (Towe et al. 1963). The ferritin in chiton blood was also isolated and characterized (Kim et al. 1986a; Nesson 1968; Stpierre et al. 1986, 1990). The ferritin isolated from chiton blood and tissues was similar in the elution patterns on agarose gel columns, aggregation behavior with blue dextran, absorption spectrum, and appearance and size in the electron microscope (Nesson 1968). Thus, ferritin in the blood may accumulate in epithelial cells of radula to supply iron for the teeth mineralization. Ferritin isolated from the chiton *A. hirtosa* blood was analyzed by ICP spectrometry. The results showed significantly low levels of phosphate content in chiton ferritin (Stpierre et al. 1990). The P:Fe ratio in chiton ferritin was 1:36, while P:Fe ratio in human ferritin and bacterioferritin is 2:1 and 1.7:1, respectively (Ueno and Watanabe 2013). The functional analysis of ferritin in *A. hirtosa* revealed that it has a M_r of 530,000, which is larger than that of mammalian ferritins and bacterioferritins. Iron content in purified ferritin lies between 1500 and 2500 atoms/mol, which is comparable to the iron content in mammalian ferritins. Note that the initial rate of iron uptake by chiton ferritin is much higher than that exhibited by mammalian ferritin (Kim et al. 1986a). These results indicate that the unique ferritin in chiton plays a key role in efficient iron accumulation for teeth mineralization.

In vitro experiments by Evans et al. showed that ferritin penetrates into the tooth interior in stage 1 and iron-demineralized stage 3 teeth (Evans et al. 1994). In that report, stage 1 teeth and iron-demineralized stage 3 teeth were infiltrated with horse spleen ferritin and analyzed after 1 week. As a result, ferritin deposited along the fibers in the teeth was detected. However, since horse spleen ferritin which has different features from chiton ferritin was used (Kim et al. 1986a), and chitin fibers may act as templates for absorption of free particles, these results need to be reexamined. Except for the results from in vitro experiments, to date, no ferritin granules have been detected inside the teeth. Kim et al. reported that the crystallinity of ferrihydrite deposited inside the teeth cusps is more significant than that of the ferritin core (Kim et al. 1989). Therefore, it is likely that iron in ferritin is released and transported inside the teeth in a soluble form and re-deposit along the fiber as ferrihydrite (Lowenstam 1981).

Shaw et al. highlighted the role of the stylus canal in elemental delivery for teeth mineralization (Shaw et al. 2009a). In that report, the iron-containing ferritin-like particles were observed in the cells located in the stylus canal. The cells in stylus canal resemble the cells in superior epithelial tissues. Both Perl's staining and EDS mapping of partially mineralized teeth revealed the presence of an iron plume between the stylus canal and iron-mineralizing core region. The tooth core region was gradually infilled with iron, phosphorous, and calcium, followed by a complete infilling of the stylus canal with these elements (Shaw et al. 2009a).

Macey and Brooker showed that iron and phosphorous were first deposited on the junction zone located between tooth cusp and stylus in *Cryptoplax striata* (Macey and Brooker 1996). Subsequently, iron started to deposit within the tooth cap region, followed by the mineralization of core region. The increasing amount of iron at the junction zone during mineralization was also observed in *A. echinata* (Brooker et al. 2003). EDS mapping of longitudinal sections of partially mineralized teeth revealed that iron deposition starts both from the outer and inner surfaces of the magnetite cusp region.

These results suggest that iron and other elements are transported to the teeth not only through epithelial cells but also through the stylus.

3.4 Summary and Perspective

Based on the studies reported to date, the formation of mineral in chiton teeth could be postulated to proceed as follows: (1) initially, iron is transported into the tooth matrix that consists of α-chitin and proteins as a soluble form (potentially as a ferric ion) through the tooth surface from the superior epithelium and from the stylus through the junction zone, (2) adsorbed hydrated iron species (on acidic proteins coating the alpha-chitin fibers) precipitate as ferrihydrite via a heterogeneous nucleation process, (3) ferrihydrite transforms to magnetite via a solid-state transformation (potentially induced by a ferrous iron ion absorption to ferrihydrite particles) and magnetite crystals grow along the chitin fiber until the gap between particles and chitin fibers is completely filled, and (4) finally, the tooth core region is filled with apatite or iron phosphate to complete the tooth formation (Fig. 3.10). The proteins identified by proteomic analysis of chiton teeth may be involved in the regulation of the above processes. Further molecular biological analysis including genome analysis and gene expression profiles will improve our understanding about the molecular mechanisms of chiton teeth formation. A multidisciplinary approach, using material science as well as molecular biology, is key to understanding chiton teeth formation.

By mimicking the strategies of mineralization within chiton teeth, we could develop new technologies for controlling the precise patterning of magnetic crystals onto substrate that could be applied for the sensing or high-density magnetic storage. Furthermore, magnetic properties of these materials could be manipulated by controlling the morphology of magnetic crystals. Beyond magnetite, an entire library of nanomaterials (i.e., metals, metal oxides, metal nitrides, metal carbides, etc.) could be fabricated using bioinspired processing, taking control lessons from

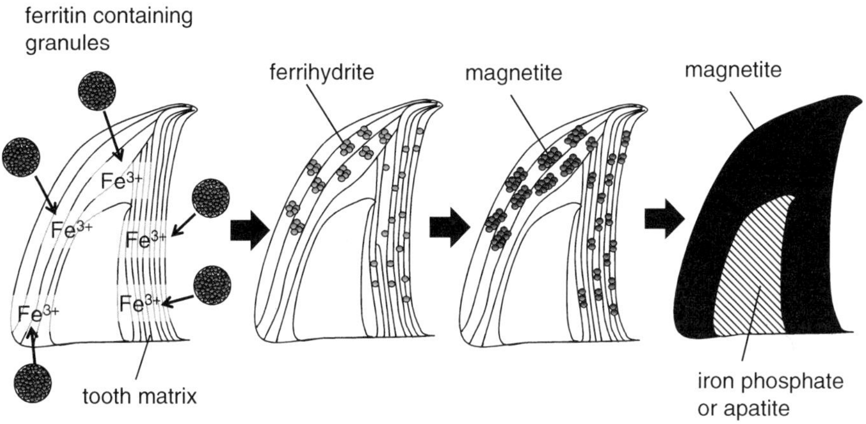

Fig. 3.10 Scheme of the hypothesized mechanism of mineral formation in chiton teeth

the chiton's precision fabrication. Furthermore, these materials, which would be made under physiological conditions, could be easily functionalized with biomolecules, enabling multifunctionality.

References

Amemiya Y, Arakaki A, Staniland SS, Tanaka T, Matsunaga T (2007) Controlled formation of magnetite crystal by partial oxidation of ferrous hydroxide in the presence of recombinant magnetotactic bacterial protein Mms6. Biomaterials 28(35):5381–5389

Arakaki A, Webb J, Matsunaga T (2003) A novel protein tightly bound to bacterial magnetic particles in Magnetospirillum magneticum strain AMB-1. J Biol Chem 278(10):8745–8750

Arakaki A, Masuda F, Amemiya Y, Tanaka T, Matsunaga T (2010) Control of the morphology and size of magnetite particles with peptides mimicking the Mms6 protein from magnetotactic bacteria. J Colloid Interface Sci 343(1):65–70

Blakemore R (1975) Magnetotactic bacteria. Science 190:377–379

Brooker LR, Lee AP, Macey DJ, van Bronswijk W, Webb J (2003) Multiple-front iron-mineralisation in chiton teeth, (Acanthopleura echinata: Mollusca: Polyplacophora). Mar Biol 142(3):447–454

Cha JN, Shimizu K, Zhou Y, Christiansen SC, Chmelka BF, Stucky GD, Morse DE (1999) Silicatein filaments and subunits from a marine sponge direct the polymerization of silica and silicones in vitro. Proc Natl Acad Sci U S A 96(2):361–365

Choi HJ, Kang SW, Yang CH, Rhee SG, Ryu SE (1998) Crystal structure of a novel human peroxidase enzyme at 2.0 angstrom resolution. Nat Struct Biol 5(5):400–406

Cornell RM, Schwertmann U (2003) The iron oxides. Wiley-VCH Verlag GmbH & Co. KGaA. Weinheim, Germany

Evans LA, Macey DJ, Webb J (1990) Characterization and structural organization of the organic matrix of the radula teeth of the chiton Acanthopleura-hirtosa. Philos Trans R Soc Lond Ser B Biol Sci 329(1252):87–96

Evans LA, Macey DJ, Webb J (1991) Distribution and composition of matrix protein in the radula teeth of the chiton Acanthopleura-hirtosa. Mar Biol 109(2):281–286

Evans LA, Macey DJ, Webb J (1992) Calcium biomineralization in the radular teeth of the chiton, Acanthopleura-hirtosa. Calcif Tissue Int 51(1):78–82
Evans LA, Macey DJ, Webb J (1994) Matrix heterogeneity in the radular teeth of the chiton Acanthopleura-hirtosa. Acta Zool 75(1):75–79
Fassbinder JW, Stanjek H, Vali H (1990) Occurrence of magnetic bacteria in soil. Nature 343(6254):161–163
Gordon LM, Joester D (2011) Nanoscale chemical tomography of buried organic-inorganic interfaces in the chiton tooth. Nature 469(7329):194–197
Gordon LM, Roman JK, Everly RM, Cohen MJ, Wilker JJ, Joester D (2014) Selective formation of metastable ferrihydrite in the chiton tooth. Angew Chem Int Ed 53(43):11506–11509
Gould JL, Kirschvink JL, Deffeyes KS (1978) Bees have magnetic remanence. Science 201(4360):1026–1028
Grunenfelder LK, de Obaldia EE, Wang QQ, Li DS, Weden B, Salinas C, Wuhrer R, Zavattieri P, Kisailus D (2014) Stress and damage mitigation from oriented nanostructures within the radular teeth of *Cryptochiton stelleri*. Adv Funct Mater 24(39):6093–6104
Jansen E, Kyek A, Schafer W, Schwertmann U (2002) The structure of six-line ferrihydrite. Appl Phys A Mater Sci Process 74:S1004–S1006
Kim KS, Webb J, Macey DJ (1986a) Properties and role of ferritin in the hemolymph of the chiton Clavarizona-hirtosa. Biochim Biophys Acta 884(3):387–394
Kim KS, Webb J, Macey DJ, Cohen DD (1986b) Compositional changes during biomineralization of the radula of the chiton Clavarizona-hirtosa. J Inorg Biochem 28(2–3):337–345
Kim KS, Macey DJ, Webb J, Mann S (1989) Iron mineralization in the radula teeth of the chiton Acanthopleura-hirtosa. Proc R Soc B Biol Sci 237(1288):335-+
Kirschvink JL, Lowenstam HA (1979) Mineralization and magnetization of chiton teeth – paleomagnetic, sedimentologic, and biologic implications of organic magnetite. Earth Planet Sci Lett 44(2):193–204
Kirschvink JL, Walker MM, Chang S-B, Dizon AE, Peterson KA (1985) Chains of single-domain magnetite particles in chinook salmon, *Oncorhynchus tshawytscha*. J Comp Physiol A 157:375–381
Lee AP, Webb J, Macey DJ, van Bronswijk W, Savarese AR, de Witt GC (1998) In situ Raman spectroscopic studies of the teeth of the chiton Acanthopleura hirtosa. J Biol Inorg Chem 3(6):614–619
Lee AP, Brooker LR, Macey DJ, van Bronswijk W, Webb J (2000) Apatite mineralization in teeth of the chiton Acanthopleura echinata. Calcif Tissue Int 67(5):408–415
Lee AP, Brooker LR, Macey DJ, Webb J, van Bronswijk W (2003) A new biomineral identified in the cores of teeth from the chiton Plaxiphora albida. J Biol Inorg Chem 8(3):256–262
Lowenstam HA (1962) Magnetite in denticle capping in recent chitons. Geol Soc Am Bull 73:435–438
Lowenstam HA (1967) Lepidocrocite, an apatite mineral, and magnetic in teeth of chitons (Polyplacophora). Science 156(3780):1373–1375
Lowenstam HA (1981) Minerals formed by organisms. Science 211(4487):1126–1131
Lowenstam HA, Weiner S (1985) Transformation of amorphous calcium phosphate to crystalline dahillite in the radular teeth of chitons. Science 227(4682):51–53
Macey DJ, Brooker LR (1996) The junction zone: initial site of mineralization in radula teeth of the chiton Cryptoplax striata (Mollusca: Polyplacophora). J Morphol 230(1):33–42
Macey DJ, Brooker LR, Webb J, StPierre TG (1996) Structural organisation of the cusps of the radular teeth of the chiton Plaxiphora albida. Acta Zool 77(4):287–294
Maher BA, Taylor RM (1988) Formation of ultrafine-grained magnetite in soils. Nature 336(6197):368–370
Matsumura S, Kajiyama S, Nishimura T, Kato T (2015) Formation of helically structured chitin/CaCO3 hybrids through an approach inspired by the biomineralization processes of crustacean cuticles. Small 11(38):5127–5133

Mizota M, Maeda Y (1986) Magnetite in the radular teeth of chitons. Hyperfine Interact 29(1–4):1423–1426
Nemoto M, Wang QQ, Li DS, Pan SQ, Matsunaga T, Kisailus D (2012) Proteomic analysis from the mineralized radular teeth of the giant Pacific chiton, *Cryptochiton stelleri* (Mollusca). Proteomics 12(18):2890–2894
Nesson MH (1968) Studies on radula tooth mineralization in the polyplacophora. California Institute of Technology, Pasadena
Nesson MH, Lowenstam HA (1985) Biomineralization processes of the radula teeth of chitons. In: Kirschvink JL, Jones DS, MacFadden BJ (eds) Magnetite biomineralization and magnetoreception in organisms. Springer, New York, pp 333–363
Okoshi K, Ishii T (1996) Concentration of elements in the radular teeth of limpets, chitons, and other marine mollusks. J Mar Biotechnol 3:252–257
Pierre TGS, Evans LA, Webb J (1992) Non-stoichiometric magnetite and maghemite in the mature teeth of the chiton Acanthopleura hirtosa. Hyperfine Interact 71(1):1275–1278
Safavi-Hemami H, Young ND, Williamson NA, Purcell AW (2010) Proteomic interrogation of venom delivery in marine cone snails: novel insights into the role of the venom bulb. J Proteome Res 9(11):5610–5619
Saunders M, Kong C, Shaw JA, Macey DJ, Clode PL (2009) Characterization of biominerals in the radula teeth of the chiton, Acanthopleura hirtosa. J Struct Biol 167(1):55–61
Shaw JA, Macey DJ, Brooker LR (2008) Radula synthesis by three species of iron mineralizing molluscs: production rate and elemental demand. J Mar Biol Assoc U K 88(3):597–601
Shaw JA, Macey DJ, Brooker LR, Stockdale EJ, Saunders M, Clode PL (2009a) The Chiton stylus canal: an element delivery pathway for tooth cusp biomineralization. J Morphol 270(5):588–600
Shaw JA, Macey DJ, Brooker LR, Stockdale EJ, Saunders M, Clode PL (2009b) Ultrastructure of the epithelial cells associated with tooth biomineralization in the chiton Acanthopleura hirtosa. Microsc Microanal 15(2):154–165
Stpierre TG, Dickson DPE, Webb J, Kim KS, Macey DJ, Mann S (1986) Some magnetic-properties of the cores of various ferritins. Hyperfine Interact 29(1–4):1427–1430
Stpierre TG, Kim KS, Webb J, Mann S, Dickson DPE (1990) Biomineralization of iron – mossbauer-spectroscopy and electron-microscopy of ferritin cores from the chiton Acanthopleura-hirtosa and the limpet patella-laticostata. Inorg Chem 29(10):1870–1874
Tanaka M, Mazuyama E, Arakaki A, Matsunaga T (2011) MMS6 protein regulates crystal morphology during nano-sized magnetite biomineralization in vivo. J Biol Chem 286(8):6386–6392
Towe KM, Lowenstam HA (1967) Ultrastructure and development of iron mineralization in the radular teeth of *Cryptochiton stelleri* (Mollusca). J Ultrastruct Res 17(1):1–13
Towe KM, Lowenstam HA, Nesson MH (1963) Invertebrate ferritin: occurrence in mollusca. Science 142(3588):63–64
Tronc E, Belleville P, Jolivet JP, Livage J (1992) Transformation of ferric hydroxide into spinel by FE(II) adsorption. Langmuir 8(1):313–319
Ueno T, Watanabe YE (2013) Coordination chemistry in protein cages: principles, design, and applications. Wiley, Hoboken
Walcott C, Gould JL, Kirschvink JL (1979) Pigeons have magnets. Science 205(4410):1027–1029
Wang P, Heitman J (2005) The cyclophilins. Genome Biol 6(7):226
Wang QQ, Nemoto M, Li DS, Weaver JC, Weden B, Stegemeier J, Bozhilov KN, Wood LR, Milliron GW, Kim CS, DiMasi E, Kisailus D (2013) Phase transformations and structural developments in the radular teeth of *Cryptochiton Stelleri*. Adv Funct Mater 23(23):2908–2917
Weaver JC, Wang Q, Miserez A, Tantuccio A, Stromberg R, Bozhilov KN, Maxwell P, Nay R, Heier ST, DiMasi E, Kisailus D (2010) Analysis of an ultra hard magnetic biomineral in chiton radular teeth. Mater Today 13:42–52

Part II
Biological Templating of Magnetic Materials for Medical and Device Applications

Chapter 4
Bioengineering and Biotechnological Applications of Bacterial Magnetic Particles

Tomoko Yoshino, Tadashi Matsunaga, and Tsuyoshi Tanaka

Abstract Magnetotactic bacteria synthesize nano-sized magnetic particles in the cells. The bacterial magnetic particles have a core of magnetite (Fe_3O_4) and are surrounded by a lipid bilayer membrane containing a number of proteins, referred to as magnetosome. Since the bioengineering methodology for magnetotactic bacteria was established, expression of a wide range of functional proteins onto magnetic particles has successfully been performed in which the native proteins in the lipid membrane can serve as anchors for the protein display. The expression system, here we call "magnetosome-display system," has enabled us to reduce the cost of production of protein-magnetic particle complexes. These advantages lead to creation of a variety of magnetic particles displaying functional proteins, including membrane proteins and disulfide-bonded proteins, which the basic approach cannot match. This review provides an overview of the developmental status of magnetic particles in the field of bioassays, summarizes magnetosome display system by magnetotactic bacteria, and discusses their usefulness and prospects in the medical and environmental fields. The novel system has shown considerable promise for improving the display efficiency of the difficult-to-express proteins and thus is expected to contribute to further development of functional magnetic particles toward biotechnological applications.

4.1 Introduction

In the medical/biological field, the purification of a from mixture is an important process. When the target is a biological sample such as cells, protein, or DNA, purity of product and efficiency of isolation are required. Magnetic particles are easily and rapidly separated and recovered using an external magnetic field, thereby

---

T. Yoshino · T. Matsunaga · T. Tanaka (✉)
Division of Biotechnology and Life Science, Institute of Engineering, Tokyo University of Agriculture and Technology, Koganei, Tokyo, Japan
e-mail: tsuyo@cc.tuat.ac.jp

T. Matsunaga et al. (eds.), *Biological Magnetic Materials and Applications*,
https://doi.org/10.1007/978-981-10-8069-2_4

facilitating their use for isolation and detection of biological samples. Furthermore, progress has been made in applying magnetic particles to drug delivery systems and cancer thermotherapy (magnetic hyperthermia) utilizing magnetic properties. With this broadening of application, the properties required of magnetic particles have been diversifying. Consequently, there are many studies on modifying the surface of magnetic particles using polymers or proteins, altering magnetic properties through doping with various metals and other methods. Typically, synthesizing such magnetic particles requires high temperatures and pressures, and producing a uniform particle size requires rigorous control. In addition, most magnetic particles used in the medical and biological fields need surface modification, which is a complex process requiring multiple steps.

The magnetic particle-synthesizing magnetotactic bacteria, which we introduce in this review, perform biosynthesis under physiological conditions of ordinary temperatures and pressures within each bacterium in a tube-shaped organelle called a magnetosome. This organelle has single-axis-structured nanoparticles (50–100 nm in diameter) with a core of magnetite (Fe_3O_4) crystals, each of which is covered by a lipid bilayer membrane primarily composed of phosphatidylethanolamine. These lipid bilayer membrane-covered, magnetic particles have excellent dispersibility in aqueous solutions and can be easily recovered using magnetic separation, thus providing opportunities for use in applications such as magnetic carriers in bioassays.

In 1995, a technique was created for targeting exogenous proteins onto the surface of magnetic particles through genetic modification of magnetic bacterium *Magnetospirillum magneticum* AMB-1. The expression system, here we call "magnetosome display system," has enabled us to produce a wide range of functional proteins, therefore improving and reducing the cost of production of protein-magnetic particle complexes. This advancement paved the road toward creating a variety of protein-magnetic particle complexes based on genetic engineering. However, although increased diversity was possible for the target proteins bonded to the surface of the magnetic particles, it was not possible to use bioengineering to control physical properties such as magnetic bodies or the dispersibility of the particles. In addition, to achieve the functional expression of exogenous proteins from animal or other cells in magnetotactic bacteria, more advanced expression processes are required to create truly diverse protein-magnetic particle complexes. Thus, this review provides an overview of the developmental status of magnetic particles in the field of bioassays, summarizes the technique for production of protein-magnetic particle complexes, "magnetosome display system" by magnetotactic bacteria, and discusses their usefulness and prospects in the medical and environmental fields.

4.2 Developmental Status of Commercialized Magnetic Particles

Currently, commercially available magnetic particles vary in a number of characteristics such as particle size, shape, and material depending on the manufacturer. Commonly, their structure consists of an aggregate of iron oxide particles with a diameter of a few nanometers dispersed or embedded in a polymer such as a polysaccharide, polystyrene, silica, or agarose. Covering the iron oxide particles with the polymer prevents the magnetic particles from clumping. In addition, the size and amount of magnetization of the magnetic particles are largely proportional to the mass of the iron oxide particles dispersed or embedded in the polymer, and so the force is greater in magnetic particles containing larger iron oxide particles (μm scale), thereby enabling easier and faster separation using an external magnetic field. By contrast, smaller magnetic particles (nm scale) are inferior in separation ease and speed but are characterized by higher reaction efficiency due to their greater specific surface area.

Dynabeads® (Invitrogen, Co.) and Magnosphere™ (JSR Life Sciences Co.) are examples of magnetic particles currently available in the market. These magnetic particles have a diameter of 1–3 μm and are constructed by molding many iron oxide particles 10 nm or smaller into spheres or coating them onto a polymer core. These magnetic particles are magnetic carriers used in biological research and immunoassays. Improved methods for synthesizing magnetic particles have led to the commercial availability of nano-sized magnetic particles. FG beads® (Tamagawa Seiki Co.) have a diameter of approximately 0.2 μm and contain multiple 20 nm Fe particles embedded in poly(glycidyl methacrylate). An applied study using thalidomide bound to FG beads®, mixing them with a cell homogenate, and followed by magnetic separation was conducted in an attempt to identify proteins causing deformities from thalidomide (Ito et al. 2010). Therma-Max® (JNC Co.) are nano-sized magnetic particles approximately 0.1 μm in size developed by devising a surface coating that enables magnetic recovery (Furukawa et al. 2003). The small size of Therma-Max® nanoparticles creates a large surface area and high reactivity, but the magnetic particles themselves are only weakly magnetic; therefore, separation using magnets is difficult. To resolve this problem, the surface of the magnetic particles in Therma-Max® is modified with the thermoresponsive polymer N-isopropylacrylamide (NIPAM). NIPAM is hydrophilic when below a particular temperature but becomes hydrophobic as the temperature rises. This property of NIPAM allows the reaction to take place at room or lower temperatures. As the temperature increases, the NIPAM becomes hydrophobic causing the Therma-Max® particles to clump together, thus enabling magnetic recovery. A highly sensitive immunoassay using Therma-Max® has been reported (Nagaoka et al. 2011). As seen above, several diverse magnetic particles have been developed (Fig. 4.1). However, their usefulness is limited by problems such as the complexity of surface modification, the multiple steps for preparation of protein-magnetic particle complexes, and deactivation of proteins caused by their immobilization on the particles. Thus, there is a need to develop magnetic particles that (1) have a small diameter and (2) are capable of surface modification with a target molecule in an easy and stable manner.

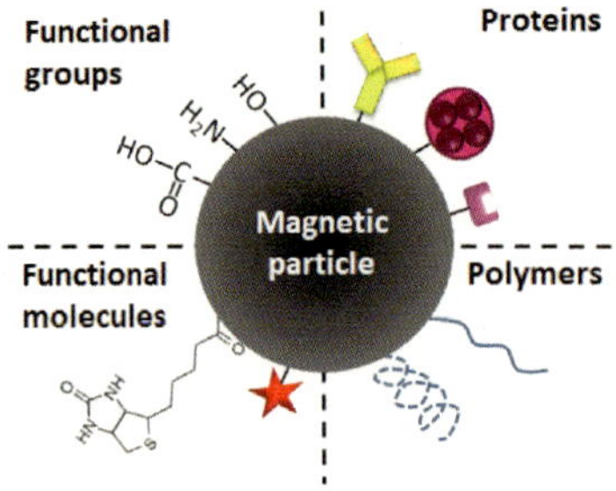

Molecules	Applications
Proteins	
Antibody	Immunoassay, cell separation
Streptavidin	Immobilization of biotinyl molecules (DNA, protein)
Protein A/G	Immobilization of antibody
Functional groups	
Carboxyl	Immobilization of various molecules
Amino	(DNA, Protein, polymer etc.)
Hydroxyl	
Functional molecules	
Biotin	Immobilization of Streptavidin
Fluorescence dye	Cell labeling, Imaging
Polymers	
PEG/PEO	Reduction of non-specific binding
NIPAM	Immunoassay

Fig. 4.1 Diverse magnetic particles and their applications

4.3 Production of Functional Magnetic Particles by "Magnetosome Display System"

4.3.1 Development of Host/Vector System

The magnetic particles in magnetotactic bacteria are in an organelle called the magnetosome and are configured from a magnetite core covered by a lipid bilayer membrane vesicle (liposome) and specific membrane proteins. The specific membrane proteins participate in magnetic particle synthesis and are important factors for exogenous protein transport and targeting during the production of protein-magnetic particle complexes. Liposomes greatly contribute to the high level of dispersibility of the magnetic particles in aqueous solutions. In addition, from the standpoint of production, liposomes are a useful platform for bringing exogenous proteins, particularly membrane proteins, into the solid phase.

The mechanism behind magnetosome formation has been detailed in numerous review articles and in this book; therefore, only the overview will be explained here. Recently, large-scale analysis techniques for genes and proteins have been coupled with the magnetosome formation mechanism, allowing research to proceed rapidly. The first magnetotactic bacterium to have its genome completely sequenced was *Magnetospirillum magneticum* AMB-1 (Matsunaga et al. 2005). The genomes of other bacteria, including *Desulfovibrio magneticus*, have been sequenced and the transcriptome and proteome analyses are in progress. In addition, the presence of numerous proteins internal and external to the membrane coating the magnetic particles has been confirmed (Arakaki et al. 2003, 2014; Tanaka et al. 2006), and analyses of the function of these proteins indicate that multiple proteins are involved in magnetic particle synthesis. Based on the results of these analyses, the proposed mechanism for magnetic particle synthesis is as follows: (1) the cytoplasmic membrane invaginates and forms liposomes arranged in a chain; (2) iron ions are taken into the liposomes; and (3) the iron concentration in the liposomes increases, causing crystallization to form Fe_3O_4 magnetite. Progress has been made in identifying and elucidating the functions of the proteins specific to magnetic particles. Among them, the proteins Mms5, 6, 7, and 13, which bind strongly to magnetite, have been shown to be involved in magnetite crystal growth. These proteins are not found in other cell fractions such as the cytoplasm; therefore, it is believed that the proteins can be specifically targeted onto magnetic particles. The Mms13 protein, which among the Mms series has a particularly high degree of localization, is strongly bonded to magnetite and hence is present and stable at the particle membrane even after washing with a solution containing a surfactant (Arakaki et al. 2003; Yoshino et al. 2008). As a result, it can be used as an "anchor molecule" for a variety of target proteins. We developed the elemental technology for the "magnetosome display system," which creates functional magnetic particles by transporting target proteins to the surface of the magnetic particles using Mms13 (Fig. 4.2), enabling free control of the magnetic particle surface. This novel technology is detailed below.

In our research group, we made rapid progress in a magnetotactic bacteria recombination system by using an oxygen-tolerant AMB-1 strain as a host and a plasmid from another magnetotactic MGT-1 strain as a vector to form a host-vector system. We developed pUMG, which is an expression vector fusing a plasmid from the MGT-1 strain and a pUC-series plasmid, enabling the free use of vectors that can be replicated in both coliform bacteria and magnetotactic bacteria (Okamura et al. 2003). Furthermore, we are investigating the development of a promoter to increase expression and control the expression period. By selecting the endogenous promoters Pmsp1, Pmsp3, and Pmms16, which are strongly expressed in *M. magneticum* AMB-1 and using them as the promoters for a target protein, the expression of that protein was markedly increased (Yoshino and Matsunaga 2005). The Schüler Group in 2008 also succeeded in expressing GFP in a strain of *M. gryphiswaldense* using

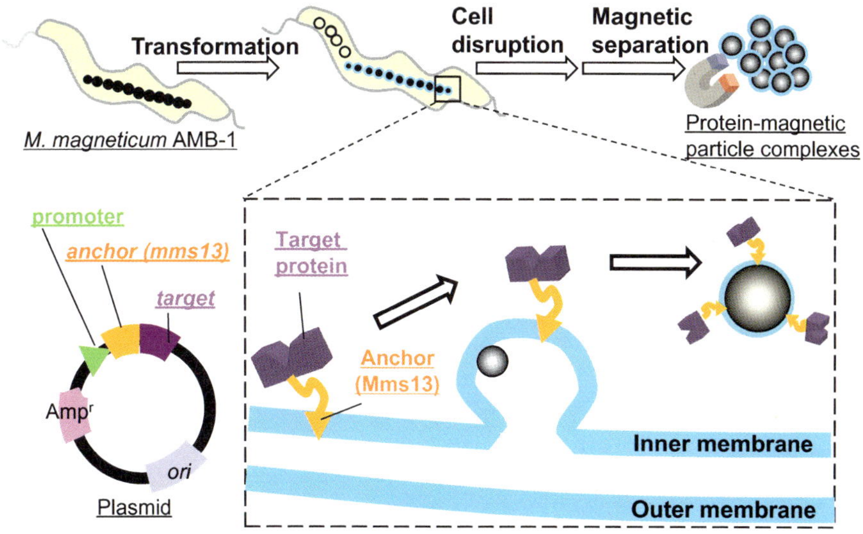

Fig. 4.2 Schematic diagram of "magnetosome-display system"

a pBBR-series vector, which was the second report of a recombination system in magnetotactic bacteria (Lang and Schüler 2008).

The author's group attempted the surface display of difficult-to-express proteins onto magnetic particles by expressing membrane proteins on liposomes. However, we encountered cases of severe growth inhibition, including the complete lack of transformants; hence, it became necessary to create an expression induction system for target proteins similar to that developed in *E. coli.* Thus, we introduced a tetracycline operator (TetO) sequence into the Pmms16 promoter region and introduced a gene expressing the tetracycline repressor (TetR) protein into the same plasmid, creating a tetracycline induction system. Normally, TetR binds to the operator, inhibiting the promoter function and turning off the expression of the target protein. In this state, by adding tetracycline (or its derivative anhydrotetracycline, ATc), which is an inducer molecule, TetR is released from TetO, and the gene expression resumes. This is very simple compared to other expression induction systems, owing to which this system was adopted early on for use with magnetotactic bacteria. We succeeded in expressing a variety of difficult-to-express proteins at the surface of magnetic particles by introducing this system into conventional protein expression vectors (Yoshino et al. 2010).

The expression induction system succeeded in displaying certain difficult-to-express proteins at the surface of magnetic particles. However, the amount expressed was very low compared to the expression of soluble proteins such as GFP or Protein A, leading us to believe that there could be decomposition in the cytoplasm or incorrect localization onto the surface of the magnetic particles. Thus, we began to modify the genome of the host magnetotactic bacteria with the aim of increasing the displayed amount of the difficult-to-express proteins. Since decomposition by proteases (protein decomposition enzymes) originating from the host cell was a possible cause for the reduced expression of the target protein, we constructed a Lon protease knockout strain (*Δlon* strain) similar to that in *E. coli* (Kanetsuki et al. 2013). We attempted to display difficult-to-express proteins at the surface of magnetic particles using the resulting magnetotactic bacteria *Δlon* strain. As a result, the expression of the target proteins on the magnetic particles was approximately three times greater than in a wild-type host. With the objective of an even greater increase in the amount displayed, we prepared an *mms13* knockout strain by inactivating the gene for Mms13 in genome DNA. In the wild-type strain, native Mms13 is localized at the surface of the magnetic particles; hence, it is believed that the display amount of target protein bound to Mms13 is limited (Fig. 4.3, wild-type). On the other hand, in the *mms13* knockout strain, only Mms13 that has bonded with the target protein is localized at the surface of the magnetic particles; hence, it was believed that a greater amount of the target proteins could be displayed on the particles (Fig. 4.3, mms13 knockout strain). As a result, the display amount of difficult-to-express protein increased by up to a factor of eight in the *mms13* knockout strain compared to that in the wild-type strain used as hosts, indicating that it was possible to increase the display of recombinant proteins (Kanetsuki et al. 2012). In this way, we demon-

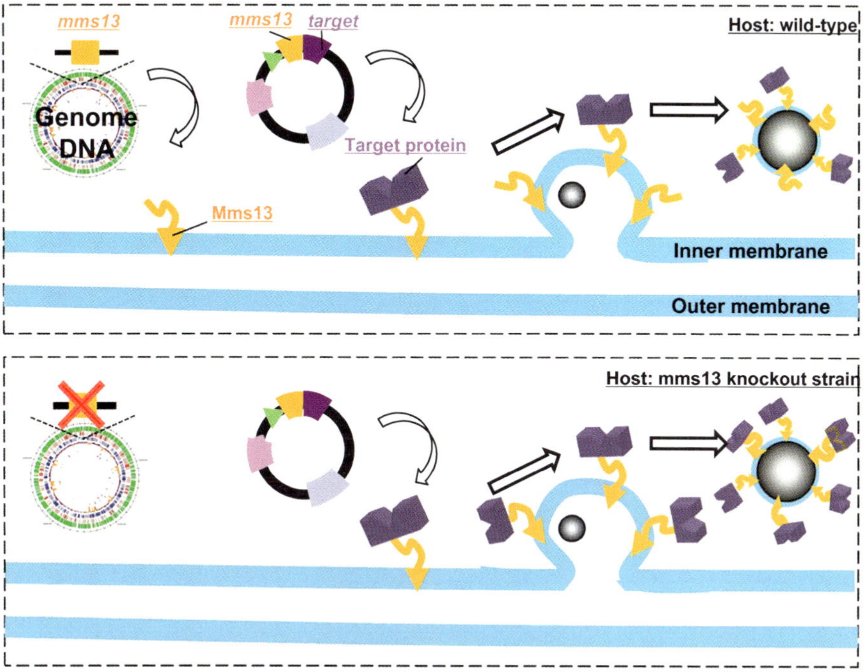

Fig. 4.3 Differences of the amount displayed on magnetic particles using wild-type or mms13 knockout strain as host

strated that the expression efficiency of difficult-to-express proteins could be greatly improved not only by selecting promoters but also by modifying the intracellular environment in the magnetotactic bacteria, which is the site of heterologous expression.

4.3.2 Novel Expression Strategies of Target Proteins onto Magnetic Particles

The functional protein-magnetic particle complexes produced by the magnetosome display system are rapidly dispersed in a reaction solution, and efficient magnetic recovery of specific substances bonded to the functional proteins is possible. Previous attempts have used the system of incorporating the gene for a target protein downstream of the gene for the anchor molecule Mms13 to efficiently transport the target to the magnetic particles. To date, various receptors and enzymes, including marker proteins such as GFP and luciferase, have been successfully expressed.

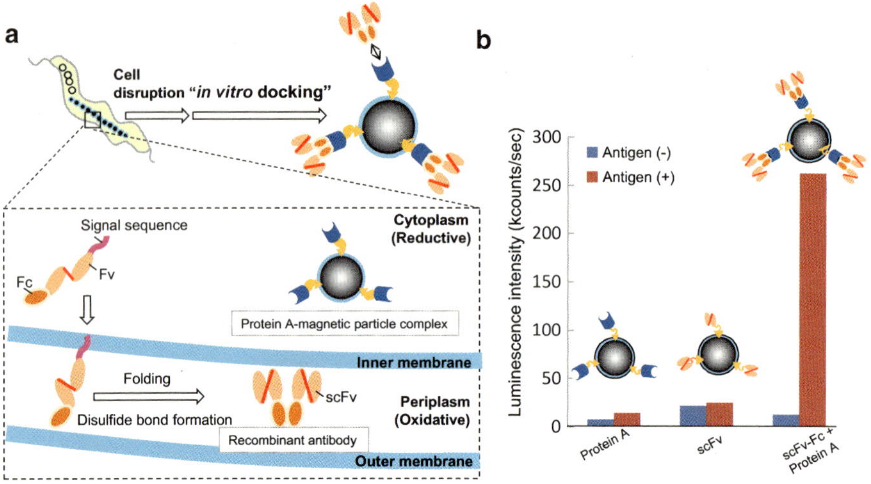

Fig. 4.4 Schematic diagram of the "*in vitro* docking" method (**a**) and antigen binding assay on magnetic particles (**b**)

As stated above, it has become possible to display a certain amount of various target proteins on the magnetic particles; however, their functionality has still not been improved. For example, attempts to express the single-chain variable fragment (scFv) on magnetic particles were successful, but no bonding with antigens could be seen (Sugamata et al. 2014). This was believed to be caused by the reducing environment in the cytoplasm in which the target protein is expressed, preventing the formation of many of the disulfide bonds in scFv, thereby destabilizing its three-dimensional structure. Thus, we took note of the fact that the periplasmic space (space between the inner and outer cell membrane) in magnetotactic bacteria is an oxidizing environment and proposed the method of forming the correct folding pattern with the formation of disulfide bonds in scFv and then immobilizing the result onto the magnetic particles. The overview of this method, termed the "*in vitro* docking," is shown in Fig. 4.4a. A recombinant antibody with a constant region (F_c) is fused to a single-chain antibody with two variable regions (F_v), and both are joined by a linker; the recombinant antibody is expressed and then transported to the periplasm. The transport to the periplasm is achieved by fusing a signal sequence (an amino acid sequence determining the localization of a protein in the cytoplasm) from a protein normally transported to the periplasm to the recombinant antibody. As a result, the recombinant antibody forms disulfide bonds in the periplasm and forms a stable dimerized structure. Meanwhile, Protein A, which strongly interacts with the constant region (F_c), is displayed at the surface of the magnetic particles. Therefore, the two structures, one in the form of magnetic particles presenting a recombinant antibody and one presenting Protein A, are produced independently in separate spaces within the cell, the periplasm and the cytoplasm, respectively. Finally, by disrupting the magnetotactic bacteria, the recombinant antibodies in the periplasm can be transferred and immobilized (docked) onto the magnetic particles.

Using this "*in vitro* docking" method, we have succeeded in affixing recombinant antibodies with antigen-specific bonding ability onto particles (Fig. 4.4b). Recombinant antibodies form dimers on the particles and have been shown to be in a state similar to the native structure of the antibodies. This can be considered evidence that disulfide bonds form within the recombinant antibodies to maintain their normal folding pattern.

In addition to the location of expression of target proteins within the cell, we attempted to control their spatial arrangement on the magnetic particles. In this method, instead of the anchor molecule binding to the target protein, a scaffold protein capable of assembling a number of proteins is bonded, enabling advanced assembling to take place. For example, let us introduce a study using the cohesin-dockerin interaction to assemble multiple enzymes to perform a cascading chain of enzymatic reactions to take place in proximity. The cohesin-dockerin interaction is characterized by high affinity (K_d value of $10^{-9 \text{ to } -11}$ M) and bond specificity. It is possible to selectively affix the desired enzyme by using multiple types of cohesin-dockerin interaction combinations. Thus, we developed a technique for using the magnetosome display method to immobilize two types of enzymes in proximity on magnetic particles. The overview of this technique is shown in Fig. 4.5a. This technique used the interaction of cohesin and dockerin from the cellulosome of *Clostridium thermocellum* and *Ruminococcus flavefaciens*. A fusion protein was expressed that combined Mms13, cohesin derived from *C. thermocellum* (Cohesin A), and cohesin derived from *R. flavefaciens* (Cohesin B) as the scaffold protein on the magnetic particles. In addition, by expressing a fusion protein joining a target protein with dockerin from *C. thermocellum* or *R. flavefaciens* (Enzyme A-Dockerin A/Enzyme B-Dockerin B), respectively, it becomes possible to affix two enzymes to the magnetic particles in the cells. This technique has been shown to be capable of localizing two types of target molecule in a 1:1 ratio in proximity to each other. Furthermore, a hydrolysis experiment of carboxymethylcellulose (CMC) was performed by affixing β-glucosidase (BG) and endoglucanase (EG) derived from *C. cellulolyticum*. Thus, magnetic particles to which both EG and BG were affixed proceeded with the CMC breakdown reaction at approximately double the rate of a mixture of magnetic particles each affixed with one or the other cellulose (Fig. 4.5b). The design of enzyme reaction cascades becomes possible by freely controlling such a magnetic particle interface (Honda et al. 2015b). In addition to these experiments, there are also attempts to express membrane proteins on magnetic particles, with high hopes for future application as magnetic carriers in a wide-ranging set of fields including drug discovery, medical care, food products, and environmental tests.

4.3.3 Improving the Material Properties of Magnetic Particles

The design of the magnetic particle interface allows the display of a variety of functional proteins through methods for expressing proteins localized on the particle membrane. There have been attempts to improve the material properties of the

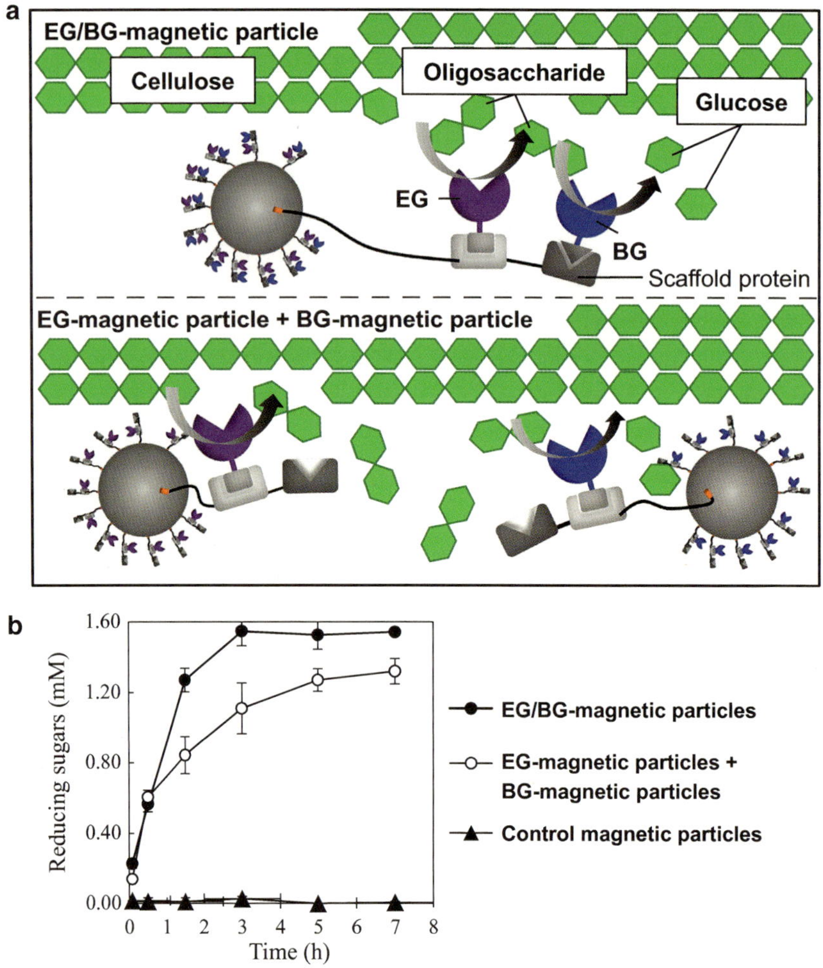

Fig. 4.5 Schematic diagram of enzymatic cascade by endoglucanase (EG) and β-glucosidase (BG)-immobilized on magnetic particles (**a**) and time course of reducing sugar contents in the supernatant of 0.5% CMC with enzyme-magnetic particle complexes (**b**)

magnetic particle themselves, as well as to challenge the limits of how far genetic engineering methods can be used to freely design magnetic particles. To date, Mms-series knockout strains in which there is strong bonding with magnetic particles have been created, and it has become clear that the morphology of the magnetic particles can be controlled. In addition, there have been attempts at surface modifications that can be controlled via responsiveness to stimuli and improvements to particle dispersibility by controlling the magnetic particle interface.

The non-specific adsorption of cells to magnetic particles can be cited as a bottleneck for improving the versatility of magnetic particles in the field of cellular medicine. To date, there have been attempts to limit the non-specific adsorption of cells to magnetic particles by means of surface modifications using synthetic polymers

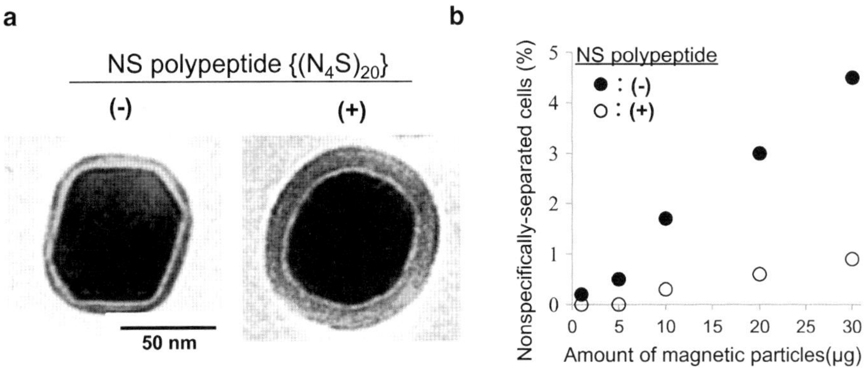

Fig. 4.6 TEM images of control magnetic particle and NS polypeptide-magnetic particle (**a**) and nonspecific binding of magnetic particles with/without NS polypeptides to the cell surfaces (**b**)

such as polyethylene glycol (PEG). PEG is a hydrophilic uncharged molecule characterized by the indispensable property of avoiding non-specific adsorption between the material and cells. With this concept in mind, artificial polypeptides were designed that mimic this property (Takahashi et al. 2010). As a result, it was discovered that non-specific adsorption to cells could be greatly limited by using genetic engineering methods to place an NS polypeptide $\{(N_4S)_{20}\}$, which is an artificial polypeptide configured from asparagine (N) and serine (S), onto magnetic particles (Fig. 4.6). NS polypeptides on magnetic particle surfaces function as a barrier to block particle aggregation and minimize non-specific adsorption of cells to the particles; they also add the ability to recognize and bind to target cells by working as a linker to display protein G on the particles. When the NS polypeptide is used in a single fusion protein as a linker to display protein G on magnetic particles, the particle acquires the capacity to specifically bind target cells and to avoid non-specific adsorption of nontarget cells.

In addition, as an application for functionalizing the surface of magnetic particles using artificial polypeptides, there has been success in developing magnetic particles with clumping/dispersion properties that can be controlled by changing the temperature by introducing a temperature-responsive polypeptide with a lower critical solution temperature (LCST) onto the particles. The greatest benefit of this method is that it is possible to design artificial proteins by means of genetic fusion with any functional protein, since each polypeptide is encoded genetically.

4.4 Developments in the Use of Magnetic Particles in the Medical Field

Magnetic iron oxide particles are widely used in medical and diagnostic applications such as magnetic resonance imaging, cell separation, drug delivery, and hyperthermia. To use these particles for biotechnological applications, the surface modification of the magnetic particle with functional molecules such as proteins, peptides, or DNA must be considered. Previously, only DNA- or antibody-immobilized magnetic particles were marketed and used in biotechnology; it was suggested that the techniques for the immobilization of enzymes or receptors were more complicated and time-consuming. However, as the methods for assembling functional proteins onto magnetic particles have become simpler and more efficient, the applications of magnetic particles have expanded. Here, the applications of proteins such as antibody, enzyme, or receptor and magnetic particle complexes are prepared by magnetosome-display system.

4.4.1 Magnetic Hyperthermia

One cancer treatment method that has been garnering attention is magnetic hyperthermia using magnetic particles. When an alternating magnetic field is imposed on magnetic particles, the orientation of the magnetic domain flips repeatedly to match that of the alternating field, resulting in the generation of heat. By accumulating magnetic particles in tumor cells in the body and applying a magnetic field, the cancer cells would be locally heated and killed, enabling the treatment of cancer with few side effects. The effect of magnetic hyperthermia is influenced by the shape and magnetic coercivity of the magnetic particles.

The magnetic particles produced by magnetotactic bacteria have excellent dispersion in aqueous solutions, as stated earlier, and hence are taken into cells without clumping and are efficiently heated. Consequently, they are garnering attention as magnetic carriers in magnetic hyperthermia (Mannucci et al. 2014). In particular, the magnetic particles produced by the AMB-1 strain can be extracted in a chain-like state via gentle cell lysis (Alphandery et al. 2012). These chains of magnetic particles reportedly have magnetic properties different from those of the pre-lysing magnetotactic bacteria or the purified magnetic particles. When an alternating magnetic field is applied to the culture medium for human mammary cancer cells to which these chains of magnetic particles have been added, the resulting cancer cell death has been shown to be more efficient than that using commercially available superparamagnetic particles (Alphandery et al. 2011). In addition, when culturing the magnetotactic bacteria strain AMB-1, adding Co^{2+} results in doping the Fe_3O_4 magnetic particle surface with Co, reportedly increasing magnetic coercivity (Staniland et al. 2008). Furthermore, there has been success in the biosynthesis of novel rod-shaped magnetic particles by controlling the expression of proteins associated with control of particle morphology. In this way, it is believed that altering the morphology and composition of the magnetic particles biosynthesized by the magnetotactic bacteria will simplify the production of magnetic materials, leading to high hopes for the development of new applications.

4.4.2 Cell Separation

As stated above, magnetic particles are used in the medical and biological fields for purifying target substances from impurities. Magnetic separation permits target cells to be isolated directly from crude samples such as the blood, bone marrow, tissue homogenates, or cultivation media. Compared to other more conventional methods of cell separation, magnetic separation may be considered a sample enrichment step for further chromatographic and electromigratory analysis. In particular, there have been attempts to isolate rare cells making up less than 1% of the total contained in the peripheral blood. The magnetic cell separation system using magnetic particles has found many uses, as it is quick and simple to use, and causes low stress to the cells. The authors have developed magnetic particles suited to the isolation of rare cells. We formed an antibody-magnetic particle complex by constructing a complex of magnetic particles and Protein A or Protein G (which can bind to antibodies) and adding an arbitrary antibody. To enrich for target cells, cell surface antigens, such as cluster of differentiation (CD) antigens, were used as markers. CD8-, CD14-, CD19-, CD20-, and CD34-positive cells were efficiently enriched from peripheral blood. The separated CD34-positive cells retained the capability of forming colonies as hematopoietic stem cells (Kuhara et al. 2004). Using this complex, it was possible to isolate CD14 cells (precursors of dendritic cells) contained in the blood at a purity of at least 93% (Takahashi et al. 2009b).

Non-specific adsorption of cells to the carrier surface is a problem with cell isolation using a solid carrier such as magnetic particles. As mentioned above, the authors newly designed a polypeptide (NS polypeptide) able to suppress non-specific adsorption. By combining this polypeptide with the magnetosome display system, magnetic particles with reduced non-specific cell adsorption were developed. These results were believed to stem from reduced interactions among magnetic particles and between magnetic particles and cells caused by the hydrophilic polypeptide. When cell isolation was performed with an antibody-bound magnetic particle, 98% cell purity was achieved (Takahashi et al. 2010).

Work has begun not only on cell isolation dependent on surface CD antigens but also on targeting other useful cells, such as killer T cells (Takahashi et al. 2009a). Recently, the authors' group has succeeded in the functional expression on magnetic particles of a complex of an antigen peptide and major histocompatibility complex (MHC) molecule, which is a cell membrane protein that recognizes T cells. The antigen peptide/MHC I molecule complex is used in the isolation of antigen-specific cytotoxic T lymphocyte (CTL) for cancer immunotherapy; however, the current method for producing the complex, which involves using an MHC inclusion body expression and reconstruction using coliform bacteria, is complicated and costly. Thus, we combined and expressed the gene coding for a single-chain polypeptide, in which the MHC I molecule heavy chain and an antigen peptide are joined with peptide linkers in the gene for the linker molecule Mms13 (Fig. 4.7a), thus increasing the complex formation rate. When cells subjected to antigen peptide stimulus from a peripheral blood mononuclear cell frac-

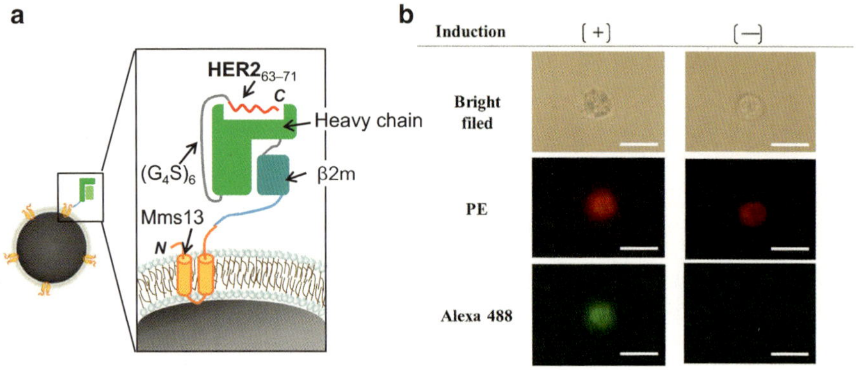

Fig. 4.7 Schematic diagram of MHC I/HER2 and magnetic particle complexes (**a**) and binding assay of the magnetic particles and HER2-induced cell. The HER263–71-induced (+) or not-induced cells (-) were stained with PE-labeled anti-CD8 antibody. Subsequently, Alexa 488-labeled magnetic particle complexes were added to the cells, and magnetic cell separation was performed (**b**). (Scale bar = 5 μm)

tion were acted upon by fluorescent-labeled MHC I/magnetic particle complexes, the ability to bond with antigen-specific CTLs was shown to be retained when using these magnetic particles (Fig. 4.7b). The MHC I molecule was not an inclusion body; thus, this was the first report of its expression in a state maintaining its functionality in prokaryotes (Honda et al. 2015a).

4.4.3 Receptor Analysis

Membrane receptors that play a role in signaling make up approximately 50% of all the targeted proteins during drug discovery. These membrane receptors are difficult to produce and must be immobilized onto a solid carrier to evaluate their function. Thus, various methods for their production and immobilization are being studied. The authors are developing bacterial magnetic particles for receptor analysis in the field of drug discovery. To date, we have succeeded in displaying a dopamine receptor (Yoshino et al. 2004), CD81 (Yoshino et al. 2010), TSHR (Sugamata et al. 2013), and chemokine receptors (CXCR4, CCR5) (Yoshino et al. 2010) on magnetic particles. Recently, we have succeeded in the functional expression on magnetic particles of Tropomyosin receptor kinase A (TrkA), which is associated with a variety of conditions such as cancer and mental disorders such as Alzheimer's disease and depression (Honda et al. 2015c). TrkA has a complex structure with a ligand-binding domain and a kinase activity domain and is a difficult-to-produce protein. The production of only the ligand-binding domain of TrkA using *E. coli* has been reported. However, compounds exhibiting physiological activity and binding to sites other than the ligand-binding domain of TrkA have been recently identified, indicating the need for a compound binding test using the entirety of TrkA. Recently, the authors' group has succeeded in the full-length TrkA in the prokaryotic host by using a magnetosome display system on magnetic particles, and TrkAs displayed on magnetic particles have been shown to retain their ligand-binding ability and kinase activity. These results indicated that detecting compounds and evaluating kinase activity could be used in a compound bonding/physiological activity evaluation using full-length TrkA.

The magnetosome display technique not only produces these recombinant proteins but also easily performs purification/refining from impurities using magnetic separation by immobilizing target molecules onto the magnetic particles. The authors' group has been creating automatic assay systems using protein-magnetic particle complexes (Matsunaga et al.. 2007), and hence we believe that it will be possible to create high-throughput compound bonding tests by combining these automatic assay systems with the protein-magnetic particle complexes that are produced.

4.5 Conclusion

As discussed above, using magnetotactic bacteria as the locus for synthesizing complexes of proteins and magnetic particles, free molecular design of the surfaces of the magnetic particles has become possible by strictly controlling the protein localization and expression period as well as the spatial distribution and amino acid sequence of the proteins. By using a synthetic biology approach to make further progress in designing and modifying biological membranes that cover the surface of the magnetic particles, a new strategy can be expected for the functional expression of these proteins. We expect that combining substance metabolism and energy metabolism with the magnetosome formation mechanism will deepen our fundamental understanding of the biosynthesis of biological magnets and broaden the utility of magnetotactic bacteria as synthetic biological hosts.

References

Alphandery E, Faure S, Seksek O, Guyot F, Chebbi I (2011) Chains of magnetosomes extracted from AMB-1 magnetotactic bacteria for application in alternative magnetic field cancer therapy. ACS Nano 5(8):6279–6296.

Alphandery E, Guyot F, Chebbi I (2012) Preparation of chains of magnetosomes, isolated from Magnetospirillum magneticum strain AMB-1 magnetotactic bacteria, yielding efficient treatment of tumors using magnetic hyperthermia. Int J Pharm 434(1–2):444–452.

Arakaki A, Webb J, Matsunaga T (2003) A novel protein tightly bound to bacterial magnetic particles in Magnetospirillum magneticum strain AMB-1. J Biol Chem 278(10):8745–8750.

Arakaki A, Yamagishi A, Fukuyo A, Tanaka M, Matsunaga T (2014) Co-ordinated functions of Mms proteins define the surface structure of cubo-octahedral magnetite crystals in magnetotactic bacteria. Mol Microbiol 93(3):554–567

Furukawa H, Shimojyo R, Ohnishi N, Fukuda H, Kondo A (2003) Affinity selection of target cells from cell surface displayed libraries: a novel procedure using thermo-responsive magnetic nanoparticles. Appl Microbiol Biotechnol 62(5–6):478–483.

Honda T, Maeda Y, Yasuda T, Tanaka T, Matsunaga T, Yoshino T (2015a) Novel designs of single-chain MHC I/peptide complex for the magnetosome display system. Protein Eng Des Sel 28:53–58

Honda T, Tanaka T, Yoshino T (2015b) Stoichiometrically controlled immobilization of multiple enzymes on magnetic nanoparticles by the magnetosome display system for efficient cellulose hydrolysis. Biomacromolecules 16(12):3863–3868

Honda T, Yasuda T, Tanaka T, Hagiwara K, Arai T, Yoshino T (2015c) Functional expression of full-length TrkA in the prokaryotic host Magnetospirillum magneticum AMB-1 by using a magnetosome display system. Appl Environ Microbiol 81:1472–1476

Ito T, Ando H, Suzuki T, Ogura T, Hotta K, Imamura Y, Yamaguchi Y, Handa, H (2010) Identification of a primary target of thalidomide teratogenicity. Science 327 (5971):1345–1350.

Kanetsuki Y, Tanaka M, Tanaka T, Matsunaga T, Yoshino T (2012) Effective expression of human proteins on bacterial magnetic particles in an anchor gene deletion mutant of *Magnetospirillum magneticum* AMB-1. Biochem Biophys Res Commun 426(1):7–11.

Kanetsuki Y, Tanaka T, Matsunaga T, Yoshino T (2013) Enhanced heterologous protein display on bacterial magnetic particles using a lon protease gene deletion mutant in *Magnetospirillum magneticum* AMB-1. J Biosci Bioeng 116(1):65–70.

Kuhara M, Takeyama H, Tanaka T, Matsunaga T (2004) Magnetic cell separation using antibody binding with protein a expressed on bacterial magnetic particles. Anal Chem 76:6207–6213

Lang C, Schüler D (2008) Expression of green fluorescent protein fused to magnetosome proteins in microaerophilic magnetotactic bacteria. Appl Environ Microbiol 74(15):4944–4953

Mannucci S, Ghin L, Conti G, Tambalo S, Lascialfari A, Orlando T, Benati D, Bernardi P, Betterle N, Bassi R, Marzola P, Sbarbati A (2014) Magnetic nanoparticles from *Magnetospirillum gryphiswaldense* increase the efficacy of thermotherapy in a model of colon carcinoma. PLoS One 9(10):e108959.

Matsunaga T, Okamura Y, Fukuda Y, Wahyudi AT, Murase Y, Takeyama H (2005) Complete genome sequence of the facultative anaerobic magnetotactic bacterium *Magnetospirillum* sp. strain AMB-1. DNA Res 12(3):157–166.

Matsunaga T, Maeda Y, Yoshino T, Takeyama H, Takahashi M, Ginya H, Asahina J, Tajima H (2007) Fully automated immunoassay for detection of prostate-specific antigen using nano-magnetic beads and micro-poly styrene bead composites, 'Beads on Beads'. Anal Chim Acta 597(2):331–339.

Nagaoka H, Sato Y, Xie X, Hata H, Eguchi M, Sakurai N, Watanabe T, Saitoh H, Kondo A, Sugita S, Ohnishi N (2011) Coupling stimuli responsive magnetic nanoparticles with antibody-antigen detection in immunoassays. Anal Chem 83(24):9197–9200.

Okamura Y, Takeyama H, Sekine T, Sakaguchi T, Wahyudi AT, Sato R, Kamiya S, Matsunaga T (2003) Design and application of a new cryptic plasmid-based shuttle vector for *Magnetospirillum magneticum*. App Environ Microbiol 69(7):4274–4277.

Staniland S, Williams W, Telling N, Van Der Laan G, Harrison A, Ward B (2008) Controlled cobalt doping of magnetosomes *in vivo*. Nat Nanotechnol 3(3):158–162.

Sugamata Y, Uchiyama R, Honda T, Tanaka T, Matsunaga T, Yoshino T (2013) Functional expression of thyroid-stimulating hormone receptor on nano-sized bacterial magnetic particles in *Magnetospirillum magneticum* AMB-1. Int J Mol Sci 14:14426–14438

Sugamata Y, Tanaka T, Matsunaga T, Yoshino T (2014) Functional expression of an scFv on bacterial magnetic particles by in vitro docking. Biochem Biophys Res Commun 445(1):1–5.

Takahashi M, Akiyama Y, Ikezumi J, Nagata T, Yoshino T, Iizuka A, Yamaguchi K, Matsunaga T (2009a) Magnetic separation of melanoma specific cytotoxic T lymphocytes from a vaccinated melanoma patient's blood using MHC/peptide complex-conjugated bacterial magnetic particles. Bioconjugate Chem 20:304–309

Takahashi M, Yoshino T, Takeyama H, Matsunaga T (2009b) Direct magnetic separation of immune cells from whole blood using bacterial magnetic particles displaying protein G. Biotechnol Prog 25(1):219–226.

Takahashi M, Yoshino T, Matsunaga T (2010) Surface modification of magnetic nanoparticles using asparagines-serine polypeptide designed to control interactions with cell surfaces. Biomaterials 31(18):4952–4957.

Tanaka M, Okamura Y, Arakaki A, Tanaka T, Takeyama H, Matsunaga T (2006) Origin of magnetosome membrane: proteomic analysis of magnetosome membrane and comparison with cytoplasmic membrane. Proteomics 6(19):5234–5247.

Yoshino T, Matsunaga T (2005) Development of efficient expression system for protein display on bacterial magnetic particles. Biochem Biophys Res Commun 338(4):1678–1681.

Yoshino T, Takahashi M, Takeyama H, Okamura Y, Kato F, Matsunaga T (2004) Assembly of G protein-coupled receptors onto nanosized bacterial magnetic particles using Mms16 as an anchor molecule. Appl Environ Microbiol 70(5):2880–2885.

Yoshino T, Hirabe H, Takahashi M, Kuhara M, Takeyama H, Matsunaga T (2008) Magnetic cell separation using nano-sized bacterial magnetic particles with reconstructed magnetosome membrane. Biotechnol Bioeng 101(3):470–477

Yoshino T, Shimojo A, Maeda Y, Matsunaga T (2010) Inducible expression of transmembrane proteins on bacterial magnetic particles in *Magnetospirillum magneticum* AMB-1. Appl Environ Microbiol 76(4):1152–1157.

Chapter 5
Protein and Peptide-Mediated Synthesis of Magnetic Nanoparticles and Arrays for Biomedicine and Future Data Storage

Andrea E. Rawlings, Rosie M. Jarrald, Lori A. Somner, and Sarah S. Staniland

Abstract Fabrication of magnetic materials with a high level of control down to the nanoscale is a current synthetic challenge. Nature is able to achieve this level of precision under ambient conditions and in aqueous solutions, by using specific biomineralisation proteins to produce highly monodisperse magnetic nanoparticles within the magnetosome organelles of magnetotactic bacteria. This chapter explores the use of such proteins outside the magnetosome, in synthetic magnetite formation reactions, where their ability to control and affect the nanoparticle products in terms of size, morphology and material purity is demonstrated. Understanding how these proteins function to achieve their activity is of particular interest, and we bring together the current literature to assess the roles of sequence and self-assembly in this process. In addition to the magnetosome-derived proteins, researchers are expanding the biological toolkit of available magnetic material mineralising proteins by using and adapting others. We investigate a number of these proteins including ferritin, heat shock protein cages and even small peptides. These can be used without modification, or they can be engineered to contain artificial binding sequences, selected via processes such as phage display. Developing new mineralising sequences allows proteins to be used with materials which are not naturally occurring, such as the platinum alloys of cobalt and iron, which have industrially desirable magnetic characteristics. The proteins/peptides covered in this chapter have the potential to aid future production of precise magnetic nanoparticles for diverse applications in the biomedical and data storage fields.

A. E. Rawlings · R. M. Jarrald · L. A. Somner · S. S. Staniland (✉)
Department of Chemistry, The University of Sheffield, Sheffield, UK
e-mail: s.s.staniland@sheffield.ac.uk

T. Matsunaga et al. (eds.), *Biological Magnetic Materials and Applications*,
https://doi.org/10.1007/978-981-10-8069-2_5

5.1 Introduction

Nanoscale inorganic materials are becoming increasingly important as the use of nanotechnology grows. More specifically, magnetic nanoparticles (MNPs) have wide-ranging uses from targeted drug delivery (Muthana et al. 2015) to ultrahigh-density data storage (Terris and Thomson 2005). Magnetite MNPs are particularly useful for biomedical applications owing to their high magnetic saturation and low toxicity, while platinum alloy MNPs such as CoPt and FePt in their $L1_0$ phase are ideal for data storage owing to their very high magnetoanisotropy and their ability to retain this at very small particle sizes. However, the reliable production of highly specific monodispersed MNP for such applications is a considerable challenge. There is therefore a critical need to develop new synthetic routes to precisely tailored MNPs.

Natural organisms carefully control the production of a vast range of inorganic minerals through biomineralisation, which is detailed in the previous chapters of this book. MNPs are biomineralised naturally by magnetotactic bacteria (MTB) that take up soluble iron ions from solution and crystallise magnetite MNPs in liposomes within their cells (called magnetosomes). Magnetosomes' size and morphology varies greatly between strains, but is exactly adhered to within each strain, demonstrating the precision that biomineralisation proteins have over the process. The mechanism of biomineralisation in MTB is the subject of chapters X and Y. The process is controlled by a unique suite of biomineralisation proteins embedded into the magnetosome membrane. Many of these proteins perform functions that we can easily replicate in a chemical test tube, such as the action of redox proteins and iron pumps. In a chemical synthesis, we can simply add iron ions and change the pH at will. However, some of these proteins are able to control the nucleation, crystallisation and morphology of the growing mineral with a very high specificity that chemists cannot currently replicate under ambient conditions, and these proteins will be the focus of this chapter.

In this chapter we explore the prospect of using nature's MTB biomineralisation proteins independently of the cell. Here we discuss using such proteins in vitro to mediate and control a simple synthetic, solution phase, chemical formation of MNPs. We then explore moving away from natural MTB proteins to synthetic proteins and peptides for the production of magnetite and also other MNPs such as CoPt & FePt. However, first we must consider the chemical synthesis of MNPs.

There are many different magnetic materials and several synthetic routes for each type, leading to a vast number of possible syntheses to discuss. For clarity and relevance to the desired applications, we will only focus on magnetite and CoPt & FePt. For both of these materials, there is a range of synthetic methods to produce them. As this chapter is concerned with simulating biological processes outside the cell, we will only focus on solution-based synthetic methods performed under conditions close to biological/ambient conditions that will be relevant for protein and peptide addition.

Magnetite (Fe_3O_4) is strongly ferrimagnetic, retains its single-domain character between approximately 30 and 80 nm (depending on the morphology) but is magnetically soft (low coercivity), with its dipole easily switched in a changing magnetic field. Magnetite is biocompatible, being present naturally in many biological organisms. It can be synthesised with a range of particle sizes and shapes, offering different magnetic characteristics. CoPt on the other hand is strongly ferromagnetic. This alloy has a chemically disordered face-centred cubic structure, also known as the A1 phase, where there is an equal probability of finding either Co or Pt in any given site. The CoPt $L1_0$ phase (Laughlin et al. 2005) can be obtained through post synthesis high-temperature (>600 °C) annealing. This is a chemically ordered face-centred tetragonal structure with alternating planes of Co and Pt resulting in a very high magnetocrystalline anisotropy energy. This is due to strong structural coupling of anisotropic atoms, giving a very high coercivity and offering single-domain character in particles down to a few nm (meaning these particles can retain their magnetisation direction in varying fields), making them ideally suited to high-density data storage (Mayes et al. 2003).

Chemical co-precipitation is the simplest method of producing both of these MNP materials by oxidising/precipitating Fe salts out of solution via increasing pH to make magnetite (under N_2) or by reducing/precipitating Co and Pt salts out of solution with $LiBEt_3H$ to form CoPt (Sun et al. 2003). This method is quick and cheap and produces large yields, but the particle quality, size and morphology are very difficult to control and typically result in a heterogeneous population. Variation in precursors, conditions and processing results in different MNP profiles. For example, with magnetite, varying either the type or method of base addition, the speciation and ratio of the iron ions (ferric/ferrous), the available oxygen in the atmosphere and the temperature can result in different sizes and shapes of particles with different iron oxides of varying crystallinity. For example, a co-precipitation of ferric and ferrous ions at room temperature under N_2 with KOH used to raise the pH results in tiny nanoparticles of poor crystallinity over a wide shape and size range (<5 nm up to micrometre scales), whereas the partial oxidation of ferrous hydroxide using KOH under N_2 at 90°C will precipitate octahedral magnetite particles approximately 20–80 nm in size. Interestingly, changing the base used can result in needle-shaped FeOOH by-products (Regazzoni et al. 1981). Although these methods allow scope to synthesise a range of materials at various sizes, the overwhelming drawback is that it is synthetically demanding to produce a monodisperse and reproducible product with respect to size and shape distribution. There are a range of high-temperature methods of producing higher quality MNP of both magnetite and CoPt with narrower dispersity; however, they require furnaces with high temperatures and pressures and/or organic solvents and toxic precursors (Sun et al. 2004). Furthermore, these conditions are difficult/dangerous to scale up to industrial manufacture and are financially and environmentally expensive.

In this chapter we investigate how we can improve the simple solution phase precipitation of these MNPs using proteins and peptides as synthesis control agents. Firstly, we consider two applications to understand and contextualise the requirements of MNP, and then we turn to nature to consider how MNPs are produced naturally by magnetotactic bacteria. We assess biomineralisation proteins and how they can be used as additives to control chemical MNP formation and consider how they achieve their function *in vitro*. We investigate the development of new material binding peptides, discovered via the technique of phage display, and how these are opening up the process of biologically controlled MNP synthesis to new materials and synthetic scenarios. Finally we show how these proteins/peptides and MNP can be coupled on planar surfaces to provide a basis for the next generation of data storage technologies.

5.2 Applications

The development of reliable, reproducible nanomaterials is essential to the evolution of dependable novel technologies, and nanoscale magnetic materials are no exception. Nanomagnets must respond to an applied field in a known and consistent manner, which requires them to have a uniform size and shape distribution. However, their individual specification with respect to size, shape, coating and degree of homogeneity of the population will vary depending on the final application. The scope of applications for MNP ranges from environmental applications such as magnetic bioremediation and contaminant sensors to biomedical diagnostics and therapies to nanotechnology such as data storage and lab-on-a-chip devices. Due to the extensive range of applications, in this chapter we will only briefly consider biomedical and data storage applications as these are directly relevant to the research described later and will enable the ultimate context and specification to be properly understood. For further, more in-depth reading on MNP applications, the reader is directed to excellent reviews describing the various applications (del Puerto et al. 2003; Lu et al. 2007; Terris et al. 2007).

5.2.1 Biomedical (Magnetite)

Magnetic nanoparticles have a wide range of applications associated with them – one of the most prominent areas of interest lies within the biomedical field. Biomedical applications of magnetite nanocrystals span the remit of both diagnostic and therapeutic areas. Diagnostics encompass any technique used to identify a marker for a certain disease; this could be a specific antigen or an increased level of a certain element/compound. Identification of biomarkers can indicate the presence or absence of a disease in the body. Research is currently being conducted into therapies that are able to treat disease and illness using magnetic nanoparticles. Examples include hyperthermia-based treatments and novel drug delivery systems. Theranostics involving magnetic nanoparticles may also prove to be important in the development of new diagnostic and therapeutic systems. Theranostics refers to the combination of diagnosis and therapy within a single treatment strategy.

When the nanoparticles are used in vivo, it is important that they are coated in a biocompatible layer to ensure it does not lead to an adverse reaction upon addition to the patient and to aid cellular uptake. Sugar or polymer coatings such as dextran or polyvinyl alcohol (PVA) may be used to increase biocompatibility, to increase colloidal stability and to provide a site for binding between the nanoparticle and the biological complex (Sangregorio et al. 1999). Monodispersity with respect to size and shape is also critical to predict the interaction within the body as different sized particles will be directed to different organs of the body. Furthermore, the MNPs must have good dispersity in solution as large sizes and/or aggregation could lead to capillary blockage and embolism.

5.2.1.1 Magnetic Separation

Magnetic separation of analytes is one instance of their use. For this example, a biological complex (such as an antibody) is tagged with a magnetic nanoparticle (such as magnetite). Following attachment the biological complex is separated from the remaining solvent or serum using magnetic separation of a fluid phase. Magnetic separation of these compounds allows the concentration of sample to be increased and used for further analysis. The magnetically tagged complex is run through a separating system where a magnetic gradient is present. The magnetic gradient immobilises the magnetic complex while the remaining fluid flows through. One disadvantage of this technique is that it is often limited by slow accumulation rates. Instead areas of high magnetism can be suspended throughout the separation vessel, immobilising the magnetic nanoparticles as they float in solution. Further optimisation of the system involves a quadrupolar set-up, where a magnetic moment is produced radially from the centre of the column (Moore et al. 2001). The labelled magnetic nanoparticles move towards the sides of the column where they become immobilised until the magnetic gradient ceases. This technique has been used to detect and diagnose the existence of parasites in the blood by magnetically tagging red blood cells (Paul et al. 1981).

5.2.1.2 MRI

Magnetic resonance imaging (MRI) is a useful diagnostic tool that utilises magnetic nanoparticles. The iron oxide particles act as contrast agents, providing increased contrast of specific tissues, organs and tumours compared to the rest of the body. The contrast provided by iron oxide nanoparticles was enhanced using magnetite nanocubes (Lee et al. 2011). MRI uses the magnetic moment of protons within the body to produce an image (e.g. of a tumour). Although the magnetic moment of a proton is extremely small, there are so many protons present that the cumulative effect allows for a detectable signal. The problem, however, is that human bodies are mainly composed of water and so protons are indiscriminately ubiquitous, providing negligible differences in contrast between different tissues. This problem is overcome by the addition of a contrast agent. A magnetic contrast agent reduces the relaxation times T_1 and T_2. T_1 is the spin-lattice relaxation time, accounting for loss of heat to the surrounding lattice. T_2 is the spin-spin relaxation time and is the measure of the time it takes for the signal to irreversibly decay; this decay occurs as excited nuclear spins in the xy plane interact with each other. Superparamagnetic iron oxide nanoparticles are commonly used contrast agents (Pankhurst et al. 2003). One example of their use is in MRI imaging of the liver. Preferential uptake of the nanoparticles in the liver is partially due to size, as MNPs with diameters of 37 nm MNPs collected more readily in the liver (Huang et al. 2010). This illustrates just how important size control of the MNPs is.

MRI can also be used to assist therapeutic strategies involving magnetic nanoparticles. The external magnetic field applied by the MRI may be used to localise the magnetic nanoparticles to one specific area of the body by directing the magnetic field gradient (Muthana et al. 2015). If this is possible, unwanted side effects of alternative therapeutics may be overcome due to the idea that the doses required for treatment will be significantly reduced and the drug will not affect the whole body – only the desired area. Problems with this method arise due to the biocompatibility of magnetite nanoparticles. Addition of coatings alters the control exhibited over their size and shape distribution during synthesis. This technique would, in theory, work alongside other biomedical applications such as hyperthermia and drug transport.

5.2.1.3 Hyperthermia

Therapeutic devices aid the treatment of a disease. Hyperthermia is one example of a possible therapeutic application of magnetite nanoparticles. Hyperthermia utilises an alternating magnetic field, which causes flipping of the electron spins within the magnetite nanoparticles. The flipping of the electron spins causes heating of the particles (Pankhurst et al. 2003). If the particles are very small, the alternating field results in the whole MNP flipping, generating heat through friction with the environment. Artificially induced hyperthermia has arisen as a possible therapy due to the ability to heat and kill tumour cells while not leading to the death of the surrounding healthy cells. Cancer cells have been found to be more susceptible to these heating effects when compared to healthy cells (Moroz et al. 2001; van der Zee 2002). Magnetic nanoparticles show promise for this application as they can be localised at the site of the diseased tissue (Hatch and Stelter 2001). The technique involves the targeting of magnetic nanoparticles to the desired tissue, followed by the application of an alternating magnetic field. The therapeutic threshold for the destruction of diseased tissue is 42 °C for 30 min. When applying this technology to patients, it is important that the alternating magnetic field is not so strong that it causes undesired effects such as arrhythmia; however, the field must be strong enough to sufficiently heat the tumour. For magnetite to be used for this application, it must first be mixed with a carrier fluid, and the size of the particle must be uniform and on the nanoscale. A monodisperse nanoparticle suspension is essential for uniform heating of the tissue.

In summary, there are various biomedical strategies that utilise magnetic nanoparticles, including (but not limited to) MRI and hyperthermia. MNPs appear to be useful in diagnostic, therapeutic and emerging theranostic applications. It is important to emphasise that the optimisation of the physical parameters of the magnetic nanocrystals is critical for their effective use in these applications.

5.2.2 Nanotechnology: Bit-Patterned Media Data Storage (CoPt/FePt)

The ideological goal of nanotechnology is to design and synthesise precise nanocomponents and assemble them accurately into increasingly complex hierarchies and designs to form intricate nanoscale devices. If this was achievable, devices such as sensors, energy accumulators and motors could be vastly miniaturised and interconnected. Data storage is an ideal model nanodevice to fulfil this aim.

5.2.2.1 Traditional Magnetic Storage

These devices consist of a thin ferromagnetic granular film, typically a Co-based alloy, deposited onto a glass or silicon surface by high-temperature sputtering (Terris and Thomson 2005). Data is recorded and read from these devices by a flying read/write head. To write, the information is transformed into a magnetic signal through an electric current, and this varying magnetic field in the write head magnetically orientates the grains contained within the magnetic film. This forms magnetic patterns in a binary code (0's and 1's) that can be read as bits of information by the read head operating in the reverse process. Since IBM introduced the first commercial magnetic hard disc in 1956, increased miniaturisation has driven the storage capacity ever higher, with today's devices having capacities in excess of 500 Gbit in^{-2} (more than 20,000,000 times the capacity of IBM's original device) (Terris and Thomson 2005). Although there are a number of semiconductor-based storage technologies such as dynamic random access memory (DRAM), magnetic devices are the most established technology (Nickolls and Dally 2010) and are significantly cheaper to produce (Terris and Thomson 2005). However, scaling the components of magnetic storage devices to ever smaller dimensions does have its limitations, and current devices are approaching their physical limits. Reducing the grain size of the granular recording media leads to domination of thermal fluctuations which induce the loss of magnetic orientation and thus the information, rendering them unsuitable for data storage as the information cannot be retained (Charap et al. 1997).

5.2.2.2 Bit-Patterned Media (BPM)

BPM is a new generation of storage media that promises to increase storage densities. It requires the formation of an array of individual magnetic islands with nanoscale dimensions onto which a bit of information is written (Terris and Thomson 2005; Terris et al. 2007). BPM has the ability to form devices with Tbit in^{-2} storage capacities (Terris et al. 2007). However, there are a number of challenges to overcome before bit-patterned media is used for ultrahigh-density data storage, including the development of a reliable and cost-effective method for the production of a recording medium containing uniform and stable magnetic islands on the nanoscale (Terris et al. 2007).

To become commercially viable, the technique used to manufacture bit-patterned storage media must be cost effective and able to be scaled up for mass production. There are three elements to this that need to be fabricated with a very tight degree of tolerance: (1) the MNP material, (2) the particle dimensions and (3) the pattern dimension.

1. *The magnetic material* is required to have a high magnetisation to ensure a signal can be clearly read, but not so high as to reorientate its neighbouring MNPs. Similarly, the MNP must be magnetically "hard", i.e. have a high-coercive force which resists reorientation when a field is reversed. This is important to ensure the information is retained. However, the force should not be so great (like a permanent magnet) as to resist information being written onto the MNP. To get the highest density of data stored in a given area, the MNPs must be as small as possible to reduce the size of each bit but not so small as to become superparamagnetic. The "magnetocrystalline anisotropy" of a material is how easily the material is to magnetise in one direction (easy axis) when compared to another (hard axis) and is a key factor in dictating its coercivity. The $L1_0$ phase of CoPt and FePt has received a lot of interest due to being two of the few magnetic materials that has high enough magnetocrystalline anisotropy to remain magnetically stable at volumes of a few nanometres in diameter (Weller et al. 2000). $L1_0$ phase 4 nm FePt particles and 18 nm CoPt have shown room temperature coercivity of 7.5 kOe (Sun et al. 2003) and 4.4 kOe (Zhang et al. 2011), respectively; however, the $L1_0$ ordered phase is difficult to achieve. Ferromagnetic CoPt alloys formed from a solid solution usually have a disordered face-centred cubic (fcc) structure, also known as the A1 phase (Aboaf et al. 1983). CoPt alloys that have a 50:50 atomic ratio are known to have distinct ferromagnetic properties and a high magnetocrystalline anisotropy energy but only after high-temperature annealing (>600 °C) (Aboaf et al. 1983) where the fcc disordered structure transforms into a face-centred tetragonal (fct) ordered $L1_0$ phase. In the A1 phase the probability that each site is occupied by either a Co or Pt atom is the same, whereas in the $L1_0$ phase two faces are occupied by one type of atom while the other face is occupied by a different atom giving it a multilayer structure that consists of alternating planes of Co and Pt which promotes the strong magnetic anisotropy and high coercivity (Laughlin et al. 2005). The high magnetocrystalline anisotropy

energy of the $L1_0$ phase means that the magnetic properties remain stable at dimensions of a few nanometres, which makes $L1_0$ CoPt or FePt ideally suited for bit-patterned media (Weller et al. 2000). However, there are problems with current methods to generate the $L1_0$ phase due to the current requirement of a high-temperature annealing step, which leads to agglomeration and sintering of the particles (Wang et al. 2014). Interestingly, however, peptide and protein-mediated synthesis of $L1_0$ CoPt and FePt at ambient temperature (see Sect. 5.4) have demonstrated the potential of a bottom-up biotemplated route to $L1_0$ Pt-alloy MNPs for data storage.

2. *Controlling the dimensions* of the MNPs arrayed on the surface is paramount to obtain the highest density storage. Ideally MNPs of 5.5 nm in diameter should be spaced 15–20 nm apart (Richter et al. 2006). However, controlling the crystallisation of this size of MNP accurately is a considerable challenge. Within this chapter we will present new possibilities of how proteins and peptide could control the size of MNP and further to this could improve this control when combined with biotemplating within a biological compartment such as a protein cage of the tailored size (see Sect. 5.4).
3. *The patterning dimensions* can be altered depending upon the technique used. There are many techniques with the capability to produce nanoscale magnetic features (Martın et al. 2003). Top-down methods such as optical lithography, although widely used in the semiconductor industry, are not currently able to achieve patterning resolutions suitable for the fabrication of bit-patterned media (Wollhofen et al. 2013). Electron-beam lithography is widely used for the production of sub-100 nm patterns but is a high-cost serial writing process that is unlikely to ever be scaled up for mass production (Driskill-Smith 2004). Low cost, versatile, soft lithographic printing methods could be financially viable but suffer from issues with reproducibility and scaling down to the nanometre length scale. The most promising top-down lithographic route looks to be interferometric lithography which is discussed in more detail in Sect. 5.5.

5.3 Mms6, MmsF and MamC for Solution Formation of Magnetite

5.3.1 Effect of Proteins In Vivo

As described in the previous chapter, the magnetosomes found in magnetotactic bacteria are protein-rich lipid vesicle organelles. The magnetosome membrane is home to a variety of different proteins which are responsible for controlling all the aspects of the magnetic nanocrystal formation, from influx of iron ions to nucleation of the iron oxide mineral and its morphological regulation. Many of these proteins reside in the lipid bilayer which surrounds the magnetosome and as such have transmembrane-spanning (TMS) regions in their sequence. Proteins such as Mms6

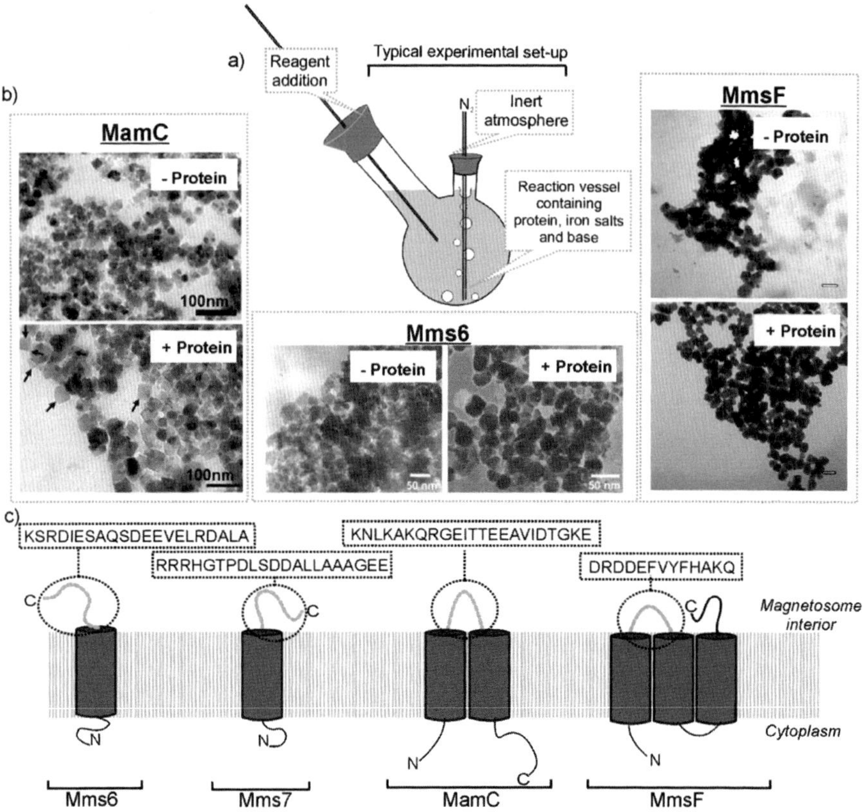

Fig. 5.1 Effect of magnetosome-derived biomineralisation proteins upon magnetite nanoparticle synthesis. (**a**) Basic reaction set-up for magnetite nanoparticle synthesis. (**b**) Transmission electron microscopy images showing the effects on nanoparticle products with and without the presence of different proteins. Proteins are Mms6 (Amemiya et al. 2007) and MmsF (Rawlings et al. 2014) from *M. magneticum* AMB-1 and MamC (Valverde-Tercedor et al. 2015) from *Magnetococcus marinus* MC-1. **c** The topology, number of transmembrane-spanning regions and potential magnetite interaction sequences of key morphology affecting proteins in the magnetosome membrane are shown. Mms6/7/F from *M. magneticum* AMB-1 and MamC from *Magnetococcus marinus* MC-1 (Images are reproduced with permissions from the references cited)

which interact with the magnetite nanocrystals directly must do so through the residues which are exposed to the interior of the magnetosome. Depending on the topology of the proteins within the membrane and the number of TMS segments that they have, these interacting regions could be N- or C-terminal domains or solvent-exposed loops connecting TMS helices (Fig. 5.1c). Mms6 and Mms7 both contain a single TMS region (Arakaki et al. 2003), and their C-terminal domain is displayed to the magnetosome interior (Fig. 5.1c). Between different strains of MTB, there are differences in naming nomenclature. Mms13, for instance, from *Magnetospirillum magneticum* AMB-1, is designated MamC in many other strains of MTB. Studies of Mms13/MamC indicate this protein adopts a two TMS structure with both N- and

C-termini exposed to the cytoplasm (Yoshino and Matsunaga 2006), meaning the loop connecting TMS helices 1 and 2 is presented to the magnetosome lumen. MmsF from *M. magneticum* AMB-1 is a triple TMS protein (Murat et al. 2012; Rawlings et al. 2014). Topological studies indicate that the N-terminal domain is located in the cytoplasm, meaning that the loop regions between TMS helices 1 and 2 must be located inside the magnetosome interior. An odd number of TMS segments can only be accomplished if the protein termini are on different faces of the membrane. Therefore the C-terminal domain of MmsF will be located in the magnetosome interior.

Mms6, Mms7 and Mms13 were all found to be tightly adhered to the magnetite nanoparticle in vivo (Arakaki et al. 2003). Having a direct binding interaction in this way suggests that these proteins may have a role in controlling the nanoparticle nucleation and/or growth. Several gene knockout and mutational studies have highlighted the importance of the genes encoding these proteins, along with MmsF, for the correct formation of the magnetite nanocrystal (Murat et al. 2012; Tanaka et al. 2010; Yamagishi et al. 2016). This makes these proteins prime targets for in vitro functional analysis.

5.3.2 *Effect of Proteins In Vitro*

The proteins present in the magnetosome membrane which are responsible for controlling and affecting the morphology of the resulting nanoparticle are also capable of exerting a similar effect when added to synthetic magnetite formation reactions (Amemiya et al. 2007; Arakaki et al. 2003, 2010; Galloway et al. 2011; Kashyap et al. 2014; Prozorov et al. 2007; Rawlings et al. 2014; Valverde-Tercedor et al. 2015; Wang et al. 2012).

The iron oxide magnetite (Fe_3O_4) comprises both ferric (Fe^{3+}) and ferrous (Fe^{2+}) iron in the stoichiometric ratio of 2:1. Many different methods exist for the synthesis of magnetite, but most of them require the presence of mixed valence iron and a high pH achieved via the addition of base (Laurent et al. 2008). Incorporating purified magnetosome-derived proteins into these types of reactions allows their effect on particle size, shape and material purity to be assessed in comparison to particles produced in their absence. The exact nature of the biological magnetite precipitation process occurring within the magnetosome has not yet been clearly resolved. However, it is likely that the current synthetic methods for studying Mms6 activity in vitro will share some similarities to the processes which take place within the magnetosome (Table 5.1).

The first recorded in vitro activity of a magnetosome-derived protein was with purified Mms6. In that example, addition of Mms6 to a simple room temperature co-precipitation (RTCP) reaction (20 μg/ml Mms6) resulted in nanoparticle products, of the majority magnetite, with a narrow size distribution and cuboidal morphology which is similar to the particle morphology found in Mms6 containing magnetosomes. Similar results to this have been achieved with reduced amounts of

Table 5.1 Summary of Mms6 *in vitro* magnetite synthesis experiments

	Protein type	MNP synthesis method	MNP size (nm)[b]	Size distribution (nm)[b]	Comments	References
Mms6 in solution	Mms6	RTCP (1:1)[a]	20–30 (1–100)	–	Tighter size distribution	Arakaki et al. (2003)
	Mms6	RTCP (1:2)[a]	22.3 (23.1)		Negligible size difference	Galloway et al. (2011)
	Mms6	POFHK	86 (234)		Smaller mean particle size	Galloway et al. (2011)
	His_6-Mms6	RTCP (1:2)[a]	21.9 (23.1)		Negligible size difference	Galloway et al. (2011)
	Mms6	POFHK	20 (27.5)	10–30 (10–40)	Smaller mean particle size, narrower size distribution	Arakaki et al. (2010)
	Mms6	POFHK	20.2 (32.4)	4.0 (9.1)[c]	Narrower size distribution, smaller mean size, cuboidal morphology	Amemiya et al. 2007
	Mms6	RTCP (1:1)[a]	21.2 (10)	8.3[c]	Larger mean size, cuboidal morphology	Amemiya et al. (2007)
Mms6 in pluronic gels	His_6-Mms6 His6-Mms6	RTCP (2:1)[a] RTCP (2:1)[a]	30 36 (4.6)	−11 (1.6)	Narrow size distribution, increased size	Prozorov et al. (2007), and Wang et al. (2012)
Mms6 peptide in solution	M6A peptide	POFHK	20–25	10–35	Exhibit spherical morphology	Arakaki et al. (2010)
	GLM6A	POFHK	20	10–30		Arakaki et al. (2010)
Surface bound Mms6	His_8-Mms6	POFHK	86 (64)	21(26)[c]	Larger mean particle size, narrower size distribution	Bird et al. (2016a)
	His_8-Mms6	POFHK	90 (69)	15(36)[c]		Bird et al. (2015)
	His_8-Mms6	POFHK	87 (60)	19 (21)[c]		Bird et al. (2016b)
Surface bound Mms6 peptide	Mms6-pep	POFHK	65 (60)	30 (21)[c]	Negligible effect	Bird et al. (2016b)
MmsF in solution	MmsF-StrepII	RTCP (1:2)	56 (45)	–	Increased size and magnetite purity	Rawlings et al. (2014)

(continued)

Table 5.1 (continued)

	Protein type	MNP synthesis method	MNP size (nm)[b]	Size distribution (nm)[b]	Comments	References
MamC in solution	MamC-His6	Free drift	30–40 (20–30)	20–80	Increased size, and morphology control	Valverde-Tercedor et al. (2015)
	MamC-truncated-His6	Free drift	30–40 (20–30)	20–80		

RTCP room temperature co-precipitation, *POFHK* partial oxidation of ferrous hydroxide with potassium hydroxide

[a]Parentheses indicate ratio of ferric to ferrous ions

[b]Parentheses indicate values for particles synthesised under identical conditions but without protein additive

[c]Standard deviation

Mms6 (Amemiya et al. 2007) (Fig. 5.1c), alternative ratios of ferrous to ferric iron in the starting reagents (Galloway et al. 2011), as well as in POFH reactions (Amemiya et al. 2007; Galloway et al. 2011), with Mms6 immobilised on planar surfaces (Bird et al. 2016b), and in pluronic gels (Prozorov et al. 2007; Wang et al. 2012). Collected together these reports indicate that the presence of Mms6 in solution typically regulates the size of the nanoparticles produced (30–20 nm) while simultaneously narrowing their size distribution, seemingly irrespective of the protein-free particle size population or reaction conditions. When used in pluronic gels, the protein increases the size of the MNP to around 30 nm, a significant increase compared to protein-free control particles (approximately 10 nm) (Prozorov et al. 2007; Wang et al. 2012). In contrast, when Mms6 is immobilised on surfaces and used in a POFH reaction, the particles produced are approximately 90 nm and much larger than the particles made under protein-free conditions (Bird et al. 2016b; Galloway et al. 2012a, b). These experiments show that using Mms6 in different forms (i.e. with fusion tags, as peptides, etc.) and in different types of experiments (in solution, in gels, on surfaces) a range of effects can be generated, potentially allowing MNP to be tailored to the particular size requirements of different precision applications.

A number of other proteins from the magnetosome have been tested using this kind of in vitro activity assay namely; MmsF, MamF, and MmxF from *Magnetospirillum magneticum* AMB-1 (Rawlings et al. 2014), and MamC from *Magnetococcus marinus* MC-1 (Valverde-Tercedor et al. 2015). Although these proteins are normally present within the magnetosome membrane, they are able to be produced in a water-soluble form (Arakaki et al. 2003; Rawlings 2016; Rawlings et al. 2014; Valverde-Tercedor et al. 2015). All of these proteins, including Mms6, show a high degree of self-assembly which results in water-soluble micelle-like structures (Amemiya et al. 2007; Feng et al. 2013; Kashyap et al. 2014; Rawlings et al. 2014; Wang et al. 2012). In these structures the hydrophobic regions of the proteins are believed to be buried within the core of the structure, and the hydrophilic, magnetite interaction regions are exposed to the surrounding aqueous environment. When purified these proteins display secondary structure, suggesting that rather than an amorphous aggregate these proteins actually adopt a specific folding architecture.

When purified, MmsF was added to synthetic RTCP reactions in amounts consistent with previous Mms6 studies. The nanoparticle products consisted of a high degree of magnetite with very little alternative iron oxide. In contrast, the addition of MamF or MmxF (both with highly similar primary and secondary structure to MmsF) shifted the balance of the nanoparticle products in favour of alternative iron oxides and away from the desired magnetite species (Rawlings et al. 2014). This suggests that potentially small residue changes can have significant effects over the type of reaction products which form.

MamC (Mms13), and a truncated form of MamC with a single transmembrane region missing, was also found to produce effects on magnetite nanoparticle synthesis (Valverde-Tercedor et al. 2015). This protein, like Mms6, has the ability to bind ferric ions in solution. When introduced into synthetic magnetite precipitation experiments performed under anaerobic "free drift" conditions over 30 days, the resulting MNP had an increased average size of 30–40 nm compared to the 20–30 nm typically obtained under control conditions. Magnetic measurements of the products confirmed the size increase.

5.3.3 Understanding the In Vitro Activity

It is evident that Mms6, MmsF and MamC are able to control, to varying degrees, the formation of magnetite MNPs when added to a chemical precipitation. It is clear that an understanding of how these proteins function is now required. The majority of the work has been conducted on Mms6 (which is the subject of a recent review (Staniland and Rawlings 2016)) as this was the first protein to be isolated and used for in vitro mediated synthesis and is also the easier (of Mms5, Mms6, Mms7 and Mms13 initially identified) to purify and work with (Arakaki et al. 2003). However, more recently, MmsF has been shown to be much easier to purify, and thus we predict more analysis will be conducted on this protein in the coming years (Rawlings et al. 2014).

To date Mms7 has yet to be purified and assessed for activity in vitro. Interestingly all four proteins, Mms6, Mms7, MmsF and MamC, have two distinct commonalities, which alludes to their function. Firstly, they all self-assemble, forming micelles or proteinosomes when purified in aqueous solution. This self-assembly is most likely due to the favourable packing of the membrane-spanning regions of the membrane proteins MamC (Valverde-Tercedor et al. 2015) and MmsF (Rawlings et al. 2014). It is believed that both the amphiphilic nature and a specific assembly sequence are responsible for organised self-assembly in the single transmembrane protein Mms6. The second common feature is the presence of multiple acidic amino acids that have been shown to bind to iron ions in the cases of Mms6 and MamC (Arakaki et al. 2003; Feng et al. 2013; Valverde-Tercedor et al. 2015; Wang et al.

2012). Here we can consider both features of self-assembly and iron ion by evaluating the work performed on Mms6 predominately and extrapolating these findings to inform suggestions about the functions of all four proteins.

5.3.3.1 Self-Assembly

Single Transmembrane Region Proteins (Mms6 and Mms7)

The amphiphilic nature of the single transmembrane proteins Mms6 and Mms7 implies they will form micelles in aqueous solution with the C-termini exposed on the surface shielding the hydrophobic N-termini within the core (Fig. 5.1c). This structure was first quantitatively investigated for Mms6 by Wang et al. who found that the micelle was between 200 and 400 kDa made up of 20–40 proteins (Wang et al. 2012). DLS measurements found Mms6 micelles were 10.2 ± 3 nm across (Wang et al. 2012), in agreement with later small-angle X-ray scattering (SAXS) experiments, which found the data fit well to a hydrophobic core of 3.9 ± 0.4 nm radius and hydrophilic corona radius of 1.1 ± 0.2 nm (at pH 3) (Zhang et al. 2015) equating to approximately 200 kDa, in agreement with the initial finding. Mms6 micelles at pH 7.5 have a slightly smaller and tighter dispersity than those at pH 3, as shown by DLS (Wang et al. 2012) and SAXS (Zhang et al. 2015), showing that varying the pH across the pI value of the protein (pI = 4.1) alters the protonation state of the acidic groups displayed on the surface, thereby affecting the structure of the micelle assembly. Interestingly, the SAXS experiment found that addition of iron caused the micelles to form higher-order structures such as discs of micelles, presumably through iron cross-linking (Zhang et al. 2015). Most recently, larger proteinous assemblies of Mms6 (approximately 10× larger) have been visualised in situ in a fluid cell TEM (Kashyap et al. 2014). It is significant that Mms6 is able to control magnetite MNP formation in vitro as in vivo, which leads to the inference that there must be self-assembly in the membrane environment in vivo similar to the aggregation seen in vitro, to retain this characteristic. We have thus suggested that Mms6 (and, due to their similarities, Mms7 also) is not monomeric in the magnetosome membrane but self-assembles to form protein rafts on the interior of the membrane, displaying a charged C-terminal surface akin to the surface of in vitro micelles (albeit of the opposite curvature). We tested this theory by enabling Mms6 to self-assemble on a surface, mimicking the membrane environment. The biomimetic Mms6 surface nucleated and controlled magnetite formation, whereas the C-terminal peptide (C20Mms6) alone could not control magnetite nanoparticle synthesis (Bird et al. 2016b). C20Mms6 is missing the N-terminal region but is still able to assemble on the surface. Thus we deduce that the nature of self-assembly in Mms6 is more specific than generic hydrophobic interactions. A GLGLGLGLG repeating amino acid sequence is notably present and conserved in both of the single transmembrane Mms protein Mms6 and Mms7 (but absent from C20Mms6). Such repeating glycine-leucine motifs are common in self-assembling scaffold proteins such as silk fibroin (Zhou et al. 2001). The large then small repeating side-chain motif lends itself to a repeating knob and holes interaction that we propose interlock with adjacent Mms6 (or even Mms7) to form a regularly packed ordered raft and thus regularly space the iron binding C-terminal sites across the magnetosome's interior membrane surface (or micelle surface in vitro). We suggest this order is critical as iron ions have been shown to bind to Mms6 (see next section), but we propose that binding in an ordered manner on the protein's surface initiates nucleation of specifically magnetite, and without this the magnetite nucleation ability is lost, as the C20Mms6 on surfaces demonstrate (Bird et al. 2016b). Supporting this hypothesis, a peptide constructed of a C6Mms6-GL repeat peptide shows much better control over particle formation than the C20Mms6, further demonstrating the importance of this region (Arakaki et al. 2010).

Multiple Transmembrane-Spanning Region Proteins (MamC and MmsF)

MamC and MmsF have two and three transmembrane-spanning regions (TMS), respectively, and therefore the sequence exposed to the interior of the magnetosome (and thus the forming magnetite crystal) is constrained in a loop configuration (Fig. 5.1c). The presence of multiple TMS dictates a dominant hydrophobicity leading to the propensity to form inclusion bodies which can be refolded to form water-soluble micelle structures similar to Mms6. This was found to be the case for MamC when it was expressed, refolded and purified (Valverde-Tercedor et al. 2015). However, unexpectedly this was not true for MmsF, with three transmembrane regions. MmsF was found to be extremely soluble (Rawlings et al. 2014). MmsF characterisation revealed it to have self-assembled into soluble "proteinosome" structures (Rawlings et al. 2014). Interestingly, MamF and MmxF (both homologues of MmsF) that are not able to control the formation of magnetite formed differently sized and more polydispersed proteinosomes (Rawlings et al. 2014), perhaps suggesting the packing within the self-assembly to again be precise and critical to magnetite formation.

5.3.3.2 Iron Binding and Proposed Magnetite Nucleation Function

To date iron binding assays have only been performed on Mms6 and MamC. The data obtained for Mms6 clearly shows the iron binding activity is dominated by interactions with the acidic aspartate and glutamate residues. Mms6 has seven acidic residues within the magnetosome lumen-exposed C-terminus, while Mms7 has five, and the loop regions on MamC and MmsF have five and four, respectively (Fig. 5.1). In addition, MamC, Mms7 and MmsF have an acidic and a basic region in their magnetosome-exposed sequences, perhaps indicating a region for iron ion binding (acidic) and a region for magnetite binding (basic). Mms6 binding studies have typically been performed on ferric (Fe^{3+}) ions, the first being demonstrated by Arakaki et al. that reported the discovery and in vitro function of Mms6 in 2003

(Arakaki et al. 2003). This paper reported a competitive radioactive ion binding assay which showed purified recombinant Mms6 binds Fe^{3+}, as well as Ca^{2+} and Mg^{2+}. Further, more quantitative ferric binding assays have been performed showing both high- and low-affinity ferric iron binding modes at low pH (Wang et al. 2012). Experiments carried out at pH 7.5 with citrate chelators to solubilise the ferric ions showed a ferric ion binding affinity of $k_d = 10^{-16}$ M, while mutants (with scrambled C-termini) showed no significant binding, demonstrating the importance of amino acid sequence in this area (Wang et al. 2012). Although there is a high affinity for ferric ion binding to Mms6 at a pH that mixed valance iron oxides such as magnetite are formed, there is no structural conformational change between the ferric binding and metal free Mms6 peptide, shown in recent 2D NMR studies, suggesting that the highly charged ferric ion could be interacting due to non-specific electrostatic interactions with the acidic residues on Mms6 (Rawlings et al. 2016). The findings from this works points to a need to assess the activity of Mms6 in situ, suggesting that tracking the chemistry quantitatively throughout the magnetite formation process is crucial to understanding the effect of the Mms6 in vitro. Delicate pH titrations were used to assess the often subtle effect of Mms6 during the magnetite precipitation process (Rawlings et al. 2016). There is no significant difference between the protein-free and Mms6-mediated precipitation at low pH (<pH 4) for a range of different ferric/ferrous ratios. This is the stage in the reaction where more insoluble ferric ions precipitate out as a ferric oxide (such as schwertmannite, hematite, or ferrihydrite depending on the conditions). This is in agreement with Wang et al. ferric binding assay which shows ferric ion binding affinities to be considerably less at pH 3 compared to pH 7 (Wang et al. 2012). It is thus clear that binding at low pH is not the main action of Mms6, being negligible when compared to the bulk precipitation. It is only after the ferric oxides have precipitated out of solution and the mixed valance iron oxides start to precipitate that a difference is seen in the Mms6-mediated titrations, showing its marked effect at this stage (Rawlings et al. 2016). This supports the hypothesis that Mms6 is most active at higher pH when the acidic groups are de-protonated and accessible for iron binding.

Interestingly Mms6 has the most marked effect in ferrous-rich ferric/ferrous ion ratios, favouring increased magnetite production (20%) with Mms6 compared to negligible amounts without protein (Rawlings et al. 2016). This suggests Mms6 directs mineralisation towards magnetite synthesis more markedly at ferrous-dominant conditions that are further from the ideal conditions for magnetite formation, effectively "pulling" minerals formed towards magnetite in preference to forming other iron oxides. Mms6 could therefore be acting as a "mineral/ferrous ion buffer" (Rawlings et al. 2016). Furthermore, this study shows that Mms6's interaction with ferrous ions is of great importance. Binding assays showed a change in structure of the Mms6 peptide (C20Mms6) with and without ferrous ions by 2D NMR (Rawlings et al. 2016). Ferrous iron produced five fold greater side-chain shifts in all the acidic groups of the DEEVE (49th–53rd amino acid) motif compared to ferric iron, signify a stronger, more specific binding of ferrous ions than ferric. Molecular modelling revealed multisite binding, suggesting Mms6 is a multi-dentate ligand for ferrous ions with the most favourable interactions between the glutamate 50 site and the adjacent carbonyl oxygen on the backbone (Rawlings et al. 2016).

The research performed to date shows Mms6, MmsF and MamC all self-assemble in aqueous solution. Furthermore, iron binding studies of Mms6 show strong specific ferrous and strong but less specific ferric binding above pH 4 at the stage in the reaction that mixed valance iron oxides nucleate, especially magnetite. We see that Mms6 binds both ferrous and ferric ions, but for a crystal of magnetite to grow, increased quantities of iron ions need to be bound, and thus there is a need to have a negatively charged iron ion binding surface that could be achieved through the protein's self-assembly. Therefore the evidence suggests that Mms6 is a magnetite nucleating protein. This has clearly been visualised by Kashyap et al., showing iron oxide nucleation on the surface of the Mms6 micelles in situ in a fluid TEM experiment (Kashyap et al. 2014). Remarkably, iron ion association with Mms6 can clearly be seen, and as the pH rises, small iron oxide particles visibly form across the micelle surface (Kashyap et al. 2014).

In vitro Mms6 forms into 10–12 nm-sized Mms6 micelles and at pH 5 and above (when the acidic groups are deprotonated) displays negatively charged surfaces for iron binding. Both ferrous and ferric ions bind, through seemingly specific (ferrous) or indiscriminately and abundantly (ferric), presumably in a 1:2 ratio, respectively. For both, the DEEVE motif is the key binding region, concentrating mixed valence iron on the surface to nucleate magnetite. It is interesting to compare this acidic motif to the other proteins. Most comparable is MmsF which has a similar four acidic residues in the form XXXZX (X = E/D, Z = V/R), albeit with the sequence reversed to Mms6. Mms7 has two double acid residue sites, which could be predicted to be iron ion binding sites. Interestingly the acidic residues on the MamC loop are well separated and appear quite different to the Mms6 binding motif, and thus more research on this specific protein to identify binding motifs involved in the iron binding mechanism is required. The spacing of bound ferric and ferrous ions is critical to nucleate magnetite, and we suggest the GL repeat sequence in Mms6 provides this by interlocking packing, precisely spacing each protein. Evidence for this hypothesis is shown in the fact that the C20Mms6 peptide surfaces (with no GL repeat) cannot nucleate magnetite as effectively (Bird et al. 2016b). Furthermore, in this same study, Mms6 and the C20Mms6 peptide only showed negligible interaction with pre-formed magnetite, showing Mms6 has no specific affinity for a magnetite surface and thus must be active in the formation process before the magnetite surface forms (Bird et al. 2016b). Again, it is indicative to see that the GL repeat motif is common to all the single transmembrane proteins (Mms5, Mms6 and Mms7) suggesting a common purpose.

We propose the action of Mms6 and the other Mms proteins (Mms7, Mms13 and MmsF) in vivo is similar to in vitro. However, instead of self-assembling into micelles or proteinosomes, within the magnetosome membrane, they would assemble as protein rafts at the membrane interior. The protein rafts could be composed of purely one type of protein or be a mixture (in the case of Mms5, Mms6 and Mms7 as they have the common GL repeat). This could be very interesting to investigate further. It is clear that Mms6 and Mms13 regulate the size of particles in vitro, while there is clear evidence Mms6 and MmsF control the iron oxide formation to be specifically magnetite. For Mms6 this size regulation is seen across a range of methods (approx. 21 nm), when nucleated on Mms6 micelles (Table 5.1). Interestingly, Mms6 assembled on planar surfaces nucleates particles approximately 90 nm in size, while MNPs within magnetosomes are 40–50 nm. The key difference between all these systems is surface curvature, moving from convex to flat to concave, respectively. We suggest this difference in degree and angle of contact between the assembled protein surface and the mineral (along with nucleation physics) is another important nucleation parameter that could be responsible for the different particle sizes (Bird et al. 2016b).

It is thought that iron is transported into magnetosomes as Fe^{2+} with subsequent partial oxidation to Fe^{3+} by specific oxidase enzymes. Interestingly Mms6 is the most influential in ferrous-rich conditions, where magnetite is more chemically challenging to achieve, but is expected to be dominant in magnetosomes. While the pH inside magnetosomes has not been determined, it must be high enough to enable magnetite to precipitate and thus the Mms proteins to be deprotonated and functional. We propose raft assemblies of Mms6 that specifically bind ferrous and ferric ions to specifically nucleate magnetite formation. Whether or not Mms6 (and by inference the other Mms proteins discussed) is a nucleating or shape controlling protein is debated in the literature. We have presented a nucleation argument for Mms6 specifically, supported by the research discussed above. However, in vivo *mms6, mms7 and mmsF* knockout mutant studies show poorly formed elongated magnetite crystals, supporting morphological controlling activity. We believe morphology and nucleation activities are coupled; if a particle is not nucleated properly, it cannot form to the correct morphology. Equally nucleation from a specific crystal plane will guide the final morphology.

5.4 Selected/Engineered Biomineralisation Proteins for Solution Formation of MNP

Naturally occurring proteins and peptides are capable of both interacting with, and affecting the formation of, different inorganic materials at the nanoscale. This makes them ideally placed to play an important role in precision control of nanomaterial fabrication in the future. However, these proteins are not without certain limitations. Firstly, many, particularly magnetosome proteins, are water insoluble (require elaborate refolding strategies) or difficult to produce in large quantities. Secondly, the magnetic materials required for future applications, such as magnetic data storage, require magnetic properties that are not available from nature, so no naturally occurring biomineralisation proteins for such materials can be exploited. Helpfully, there are emerging strategies which allow biomolecules to be developed to overcome

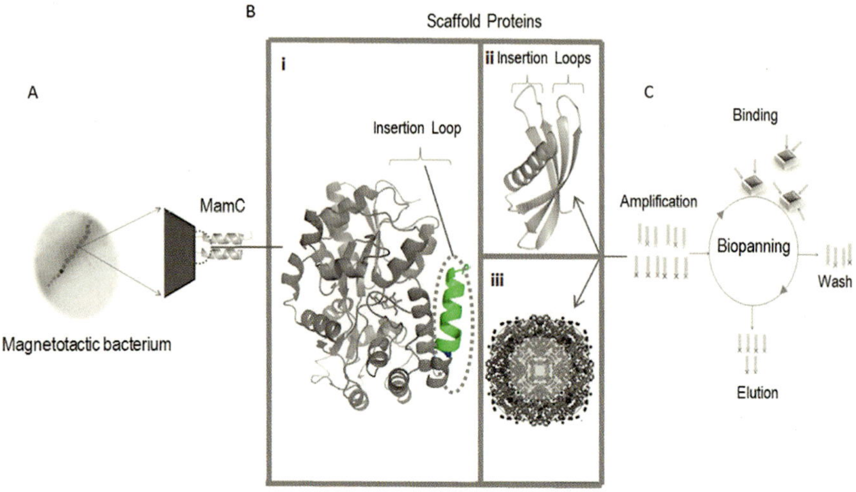

Fig. 5.2 The link between magnetotactic bacteria, phage display and scaffold proteins. (**a**) A magnified cartoon of the magnetosomes of magnetic bacteria including the transmembrane MamC biomineralisation protein. (**b**) Examples of protein scaffold additives used for magnetite synthesis in vitro. Left: MBP-MamC (PDB: 5E7U) (Nudelman et al. 2016). Top right: Adhiron protein scaffold (PDB: 4N6T) (Tiede et al. 2014). Bottom right: apoferritin (PDB: 5ERK) (Pontillo et al. 2016). **c** The biopanning cycles implemented during phage display; used in the production of MIAs and modified ferritin cages

these challenges. These include (i) modification of existing biomineralisation proteins to enhance stability, "expressibility" and solubility, (ii) repurposing of existing proteins to new mineralisation scenarios and (iii) development of entirely new peptide sequences (Fig. 5.2).

5.4.1 Engineering of Magnetosome-Associated Proteins

Many proteins can be modified and adapted to make them more amenable to certain characterisation or applications, and the biomineralisation proteins from magnetotactic bacteria are no exception. Mms6 has been modified with the addition of a single cysteine residue at the N-terminal region (Bird et al. 2016a, b). The incorporation of this non-native thiol group to the Mms6 protein enables direct immobilisation on a gold substrate for QCMD measurements and nano-patterning techniques (see Sect. 5.5.1). It is also possible to produce Mms6 as a C-terminal fusion to the highly soluble maltose binding protein (MBP). By engineering a linker sequence between the two proteins that is cleavable by tobacco etch virus protease, Mms6 can be easily released from the fusion after expression and purification without the need for lengthy refolding strategies (Galloway et al. 2012b).

Another example is where the maltose binding protein (MBP) has been used as a protein scaffold to display peptide loops from magnetotactic bacteria. Nudelman et al. have demonstrated this by fusing the magnetite binding loop of the biomineralisation protein MamC to the C-terminus of MBP (Fig. 5.2a & bi). Two constructs were produced: one short peptide consisting of 17 amino acids (R-61 to G77) and one longer peptide containing 21 amino acid residues (L57-G77). These constructs were made to analyse the relationship between the structure of the MamC loop and its binding capability to magnetite nanoparticles. For this investigation the constructs were added to in vitro magnetite synthesis reactions. The use of the MBP display scaffold aided the discovery of two important amino acid residues for MamC binding to magnetite: Glu66 and Asp70 (Nudelman et al. 2016).

5.4.2 *Peptide and Protein Biopanning Against MNP*

5.4.2.1 Principles of Biopanning

Biopanning is a way of finding novel biomolecules (usually proteins or peptides) which interact with specific substrates (Jijakli et al. 2016; Smith 1985). In this approach, a diverse library of random peptides or proteins are created which have variable amino acid sequences. This library is engineered so that an individual sequence is expressed as a fusion to an endogenous protein exposed to the outer surface of either a cell (cell display) or bacteriophage (phage display). Using a cell or phage in this way is the crux of the biopanning technique, as both the protein and the DNA, which encodes it, are physically linked together. This pool of proteins, each displayed on a cell or virus, is then applied to the desired substrate material. Some of the library members will contain a sequence which has a natural binding affinity to the material and the others will not. By washing the substrate to remove the non-binding sequences, the positive binders are effectively purified from the pool and can be selectively eluted using, for instance, pH switching of the surrounding buffer. The cells or viruses collected in this way can then be amplified, and the sequences of the binding proteins determined via sequencing of the encoding DNA. In each successive round of binding, washing, elution and sequence amplification, the positive binders which show strong affinity to the desired substrate are selectively enriched. This results in a population of high-affinity binders to carry forward for further analysis (Fig. 5.2c).

Biopanning in this way has uncovered a number of sequences which are able to interact with a range of specific materials for either simple binding or for directed material synthesis.

5.4.2.2 Magnetite Interacting Sequences

One such example of a system used for both magnetite binding and synthesis is the Adhiron protein (Fig. 5.2bii). This is a monomeric protein scaffold that is stable against a large pH range and high temperatures. This protein scaffold expresses well in *E. coli*, and the X-ray crystal structure has been resolved to 1.75 Å (Tiede et al.

2014). The protein contains up to two loop regions. These loop regions were altered so that they contained nine randomised amino acids which enabled a phage display libraries to be constructed (Tiede et al. 2014). The Adhiron library, containing the randomised loops, was screened against magnetite nanoparticle and, more specifically, magnetite cubes (Rawlings et al. 2015). This library was used to determine whether any peptide sequences preferentially bound to the specific surface of the iron oxide. These hits were named magnetite interacting Adhirons (MIAs) (Fig. 5.2bii). The library produced was found to be enriched with basic, lysine residues (Rawlings et al. 2015). Adhiron proteins containing both a single loop and two loops showed molecular recognition towards magnetite. Furthermore, these proteins appeared to encourage the formation of cubic nanoparticles in room temperature co-precipitation (RTCP) reactions – indicating the protein holds the ability to exhibit a degree of shape control on the forming magnetite nanoparticles (Rawlings et al. 2015). It is expected that the Adhiron has a major role during the growth of the crystal. It is likely that the MIA interacts with the magnetite crystal early, during the crystal growth phase. This interaction may inhibit further growth at this face, therefore promoting growth of the other crystal faces, resulting in a cubic shape. Artificial scaffolds, such as this one, have a number of benefits over naturally occurring proteins (e.g. proteins that have been repurposed). The artificial scaffolds tend to have increased solubility, expression levels and yields of production, especially when compared to proteins such as transmembrane proteins.

In a different approach, short peptides in a phage display library were screened for binding to magnetite (Baumgartner et al. 2014). For this application, phage display was not used for the direct assessment of peptide sequences, but it was used to identify similar sequences in MTB that were likely to have magnetite biomineralisation capabilities (Baumgartner et al. 2014). This removes the need for extensive mutagenesis experiments and subsequent in vivo analysis.

5.4.2.3 CoPt/FePt Interacting Sequences

Biopanning carried out by Reiss et al. in 2004 identified four peptide sequences that show specificity for $L1_0$ FePt. These peptide sequences are HNKHLPSTQPLA, SVSVGMKPSPRP, VISNHRESSRPL and KSLSRHDHIHHH (Reiss et al. 2004). Similarly, an $L1_0$ CoPt specific peptide was identified by phage display by Klem et al in 2005 with the sequence KTHEIHSPLLHK (Klem et al. 2005). All these peptide sequences contain a large number of basic amine residues, such as lysine, which is an excellent ligand to bind to Pt. Lysine is believed to be essential in the mechanism for binding of high mobility group domain proteins to the Pt/DNA complex that is formed in cisplatin-based therapies (Mikata et al. 2001). Therefore it is likely lysine will bind to platinum during the synthesis of FePt and CoPt NPs. For FePt HNKHLPSTQPLA was identified through biopanning against FePt NPs and a thin film of $L1_0$ phase FePt; therefore this was the only peptide used in subsequent experiments. In that study FePt NPs were produced by mixing $FeCl_2$, H_2PtCl_6 and $NaBH_4$ in a 1:1:1 molar ratio under ambient conditions, with 1 ml of engineered phage present. NPs with a diameter of 4 nm were produced, and selected area diffraction patterns showed rings corresponding to (001), (110), (111) and (200). The presence of the (001) and (110) planes indicates that a proportion of the NPs are in the $L1_0$ phase (Reiss et al. 2004).

Magnetic measurements performed at 5 and 300 K showed coercivities of 1350 and 200 Oe, respectively (Reiss et al. 2004). The coercivity measured at room temperature is lower than that expected for a sample containing $L1_0$ NPs. This could be because the particles produced in the reaction are close to the SPM limit of $L1_0$ FePt or the NPs have less $L1_0$ character than predicted. As a control experiment, FePt particles were synthesised in the presence of a peptide that is not specific to FePt (SPPRNYYSSMSS) (Reiss et al. 2004). From the SAED of these particles, only one band can be seen, and this relates to the (111) plane of FePt; this plane is not indicative of $L1_0$ and therefore suggests that very little chemical ordering occurs when a non-specific peptide is used (Reiss et al. 2004).

5.4.3 Repurposing Protein Cages for MNP Synthesis

Protein cages can be used as a biotemplate in the synthesis of magnetic nanoparticles. Due to the multiple subunit nature of protein cages, modification of a single subunit will result in modification of the fully assembled cage. There are three surfaces that can be modified in a protein cage: the internal and external surfaces and the interface between the subunits. Modification of a protein cage can result in the formation of a biological template that has a high affinity for a specific material.

5.4.3.1 Unmodified Protein Cages

Ferritin

Ferritin is a naturally occurring, cage protein used for the storage and biomineralisation of inorganic minerals. The mammalian ferritin precursor, apoferritin, consists of a 24-subunit assembly with a four-helix bundle forming a dodecameric cage. The interior compartment of the protein is between 7 and 8 nm in diameter and can act as a nanoreactor, promoting the synthesis of monodisperse nanomaterials in the protein core. These nanomaterials can be used for a wide range of applications (Jutz et al. 2015). The outer shell of ferritin can also be utilised. It can be functionalised chemically or genetically. This functionalisation enables the ferritin protein shell to act as a multivalent scaffold. For example, functionalisation of the protein shell can render ferritin particles soluble in organic solvents, which can be advantageous for many applications. This has been shown by Wong et al. (1999) and Sengonul et al. (2007). Functionalisation of the outer surface of ferritin, combined with a controlled inorganic core, gives technologically useful, biocompatible complexes. The core acts as a template for the formation of monodisperse nanocrystals, with inorganic material becoming incorporated into the core through the internal channels or through reversible disassembly and reassembly of the cage. Nucleation can occur on the surface of the ferritin internal cavity; this occurrence depends on the metal ion binding capability of the exposed charged amino acids on the protein. Alternatively inorganic cores can form by directed mineralisation within the ferritin cage; this is thought to occur by forming favourable electrostatic interactions with the metal to induce crystal nuclei formation. The first mineral cores to be synthesised within ferritin were iron sulphide and magnesium oxide (Meldrum et al. 1991). Later studies showed the formation of nanoparticles which possess properties that make them useful for various applications, such as magnetic nanoparticles (i.e. magnetite). The magnetic properties of the MNP core can be altered without changing the size of the crystal. This is because the size is tightly controlled by the size of the channels and the rigidity of the protein template. Doping with other transition metal elements such as cobalt and manganese can increase the magnetic hardness of the complex. The development of clinically applicable systems was enhanced further upon the synthesis of gadolinium (III) oxide hydroxide by incubating a gadolinium nitrate solution with ferritin (Sanchez et al. 2009).

As mentioned previously, phage display libraries have been designed to identify peptide sequences that bind strongly and specifically to certain inorganic surfaces. These sequences can be used to link inorganic particles to biomolecules fused to the peptide. These peptides often exhibit stronger, more specific binding when attached to a rigid scaffold. In phage display, the phage can occupy this role; however, once the peptide has been released from the phage, it often requires attachment to another display protein – such as ferritin. A specific binding sequence for titanium was inserted into ferritin (via a DNA cassette encoding the peptide) between the first and second codons of the L-chain gene. The horse L-chain that was used consists of 174 amino acids, arranged into four long helices. The adapted ferritin cages still maintain the capability to biomineralise iron but are also capable of selectively binding other inorganic material, such as titanium, with a higher affinity than the titanium-binding peptide in the absence of the ferritin scaffold (Sano et al. 2005).

MjHsp

A well-studied example of a protein cage is the small heat shock protein cage from *Methanococcus jannaschii* (MjHsp). MjHsp assembles into an empty 24-subunit cage with exterior and interior diameters of 12 nm and 6 nm, respectively. Large 3 nm pores at the threefold axes allow exchange between the interior of the cage and the bulk solution (Kim et al. 1998). The protein cage is stable up to 70 °C and in a pH range of 5–11. MjHsp has been used to encapsulate ferrihydrite through air oxidation of Fe(II) in the presence of the protein cage; this is analogous to the ferritin reaction (Flenniken et al. 2003). These particles were imaged using TEM, and an average diameter of 9.0 nm was observed; the presence of iron was confirmed by electron energy loss spectroscopy (EELS). As a control Fe(II) was oxidised in the absence of MjHsp, resulting in a rapid precipitation of bulk ferric oxide due to unconstrained particle growth. Electron diffraction data suggested that the bulk solution was consistent with lepidocrocite and the particles synthesised in the presence of MjHsp were consistent with ferrihydrite (Flenniken et al. 2003), suggesting that the MjHsp protein cage acts as a constrained reaction vessel for the mineralisation of Fe(O)OH.

PepA

PepA is protein shell from *Streptococcus pneumoniae* that has a hollow tetrahedral structure with a 6 nm interior cavity and a 12 nm exterior diameter. The cavity at the centre has four wide channels at the faces of the tetrahedron and four narrow channels at the edges (San et al. 2013). The wide and narrow channels have a diameter of 4 nm and 1 nm, respectively. PepA has been used to mediate the growth of CoPt NPs inside the protein cavity under ambient reaction conditions (San et al. 2013). The mechanism of how the protein shell mediates NP growth is unclear (Douglas et al. 2002). However, it has been suggested that it occurs via a complementary electrostatic mechanism. In this proposed method, it is thought that Co^{2+} and Pt^{2+} diffuse in and accumulate in PepA through the channels. The positively charged precursor metals interact with the negatively charged interior of the cage resulting in the formation of nucleating intermediates. CoPt NPs then grow by further deposition at the NP seeds (Douglas et al. 2002).

CoPt NPs synthesised using this method have a size range of 1.1–2.8 nm. The 1.1 nm particles exhibit superparamagnetic behaviour, while NPs greater than 2.1 nm exhibit ferromagnetic behaviour but only at 5 K. XRD data shows diffraction peaks at (111), (200), (220) and (311) all of which are typical for CoPt. Coercivities of 775 and 800 Oe were observed for particles of 2.1 nm and 2.8 nm, respectively, and a general trend of decreased magnetisation was observed as particle size decreased (San et al. 2013).

Dps

Dps is a ferritin-like protein from the bacteria *Listeria innocua* that forms a 12-subunit cage structure with 3:2 tetrahedral symmetry (Bozzi et al. 1997). The Dps protein has some motifs that are structurally similar to ferritin and also contain a ferroxidase centre. The cage has an exterior and interior diameter of 9 and 6 nm, respectively. Ferrimagnetic Fe_3O_4 nanoparticles have been synthesised by the controlled oxidation of iron at an elevated temperature (65 °C) and pH (8.5). The samples were characterised via TEM, and a narrow size distribution of 4.5–6.0 nm was identified with an average of 500 Fe atoms per cage (Allen et al. 2002; Bozzi et al. 1997). It is also possible to synthesise Co_3O_4 nanoparticles in the Dps protein cage by substitution of Co (II) for Fe(II) (Allen et al. 2003).

5.4.3.2 Modified Protein Cages

By combining sequences obtained from phage display with protein cage architectures, magnetic nanoparticles can be synthesised inside the protein cavity, resulting in a high level of control (Fig. 5.2biii).

MjHsp

MjHsp has external and internal surfaces that can be chemically or genetically modified. The sequence for the $L1_0$ phase of CoPt peptide (KTHEIHSPLLHK) was incorporated at the N-terminus of the protein cage, which expressed on the interior (Klem et al. 2005). Reaction of Co(II) and Pt(II) salts under reducing conditions led to the formation of CoPt NPs inside the protein cages. Nanoparticles synthesised in this reaction were analysed by TEM and have a mean diameter of 6.5 nm, which corresponds to the interior diameter of the protein cage (Klem et al. 2005).

Magnetic measurements of MNPs produced in the presence of the modified protein cage showed a coercivity of 150 Oe at room temperature. No hysteresis was observed at room temperature for NPs mineralised in the presence of the wild-type protein cage; instead a superparamagnetic component was identified, and this is consistent with the absence of $L1_0$ structure (Klem et al. 2005). High-temperature annealing (650 °C) leads to a coercivity of 650 Oe for NPs mineralised with the modified protein cage. This high-temperature annealing step leads to the alignment of the c-axis and therefore increased magnetism (Klem et al. 2005). However, high-temperature annealing can cause agglomeration of adjacent particles, which increases particle size and size distributions, therefore removing the advantages of the biotemplating process.

CPMV

The cowpea mosaic virus (CPMV) is an example of a virus that has been used for the formation of MNPs. CPMV is 28 nm in diameter, and the surface of the virus particle has exposed amino acids suitable for the attachment of proteins (Love et al. 2014). For example, there are exposed amines (lysine), carboxylates (aspartic acid and glutamic acid) and hydroxyl (tyrosine) groups. Exposed functional groups have been used to conjugate surface amines to a peptide that directs specific mineralisation processes. Peptides specific for the mineralisation of FePt (HNKHLPSTQPLA) and CoPt (CNAGDHANC) were bound to CPMV through chemical coupling of exposed surface amine groups (Aljabali et al. 2011a, b). The peptide that has been used in this study for the mineralisation of CoPt is different from that in previous studies (KTHEIHSPLLHK); this is because it is not specific for the $L1_0$ phase of CoPt but just to the mineralisation of CoPt (Klem et al. 2005). By attaching these peptides to the surface of CPMV, CoPt- and FePt-coated nanospheres with a diameter of 32 nm could be formed at room temperature (Aljabali et al. 2011b).

5.5 Patterned Surface Arrays of MNP Using Proteins and Peptides

In the previous section of this chapter, we have clearly identified active peptide regions of both Mms proteins and biopanned selected MIA (magnetite) proteins and explained how they may control nucleation and crystal growth/morphology. These proteins and "easier to produce" mimics have been used extensively and reliably as additives to control magnetite MNP formation in "green" chemical precipitation. Additionally, this has been expanded into the synthesis of other MNP materials such as CoPt using active peptide regions, and these can be displayed on cage protein scaffolds. Furthermore, we have explored the potential of such a route for nanotechnological applications, namely, high-density data storage. In this section we expand on this in vitro methodology to develop more complex architectures, producing MNPs attached to surfaces and the patterning of these surfaces to make defined arrays of MNPs on surfaces (Fig. 5.3). In doing so, we are expanding the function

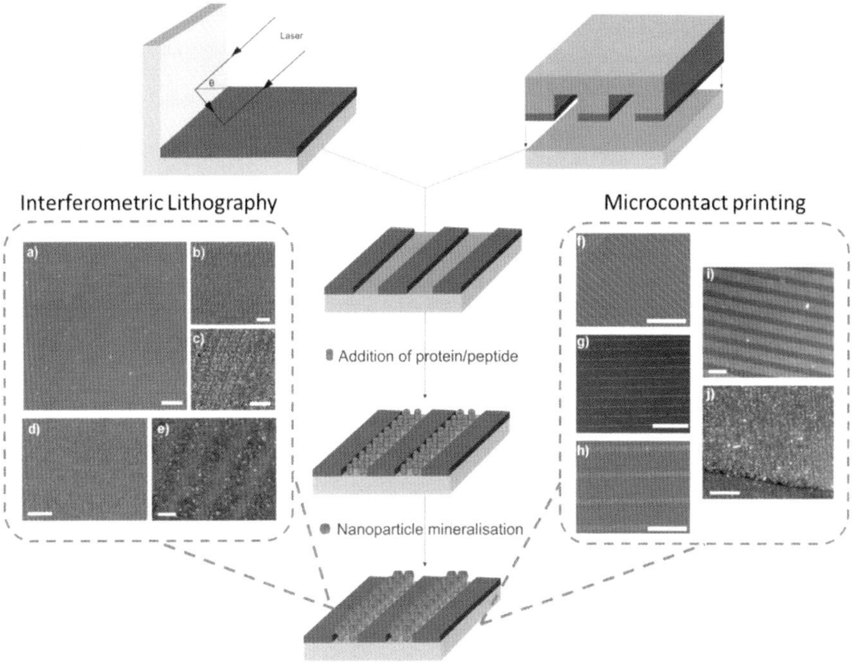

Fig. 5.3 Schematic outlining the two methods of patterning discussed, micro-contact printing and interferometric lithography. Scanning electron micrograph images show patterns produced using these two methods: (**a–c**) Mms6 biotemplated MNP arrays patterned by IL (Bird et al. 2016a), (**d**, **e**) cys-CoPt peptide biotemplated MNP arrays patterned by IL (Galloway et al. 2016) and (**f–h**) μCP cobalt-doped magnetite arrays (Bird et al. 2015) and (**i**, **j**) μCP patterned arrays of CoPt (Galloway et al. 2016). Scale bars represent: (**a**) 2 μm, (**b**) 1 μm, (**c**) 500 nm, (**d**) 20 μm, (**e**) 200 nm, (**f**) 100 μm, (**g**) and (**h**) 25 μm, (**i**) 20 μm and (**j**) 2 μm (Images are reproduced with permission from The Royal Society of Chemistry)

of the protein, giving it a dual purpose. This is because the protein precisely controls the particles' composition, size and/or morphology while also locating and immobilising the particle on the surface to facilitate patterning.

The first benefit of mineralising MNPs on proteins displayed on surfaces is that this 2D arrangement more closely mimics the 2D native lipid membrane environment of the protein in vivo. This has been used as a model to understand how the protein functions. However, the key motivation behind producing protein-templated MNPs in arrays on surfaces is the potential to develop a "green" biosynthetic method to produce precise and tailored nanodevices, specifically high-density data storage media (see Sect. 5.4.2). To this end, bio-additives to control the formation $L1_0$ Pt-alloy MNPs are particularly relevant as ideal information storage nanoparticles. Magnetite and $L1_0$ CoPt MNPs biotemplated on surfaces are therefore discussed in this section.

The methodology can be broken down into three distinct components:

1. *The type of MNP to be patterned and thus the choice of protein or peptide*. In this section we will review the formation of magnetite MNPs on surfaces using Mms6, ferritin cages, CoPt in ferritin cages and the CoPt-binding peptide KTHEIHSPLLHK previously discussed.
2. *The method of attachment to the surface*. Patterning a self-assembled monolayer (SAM) with contrasting functional groups is a well-established field of surface chemistry. Briefly, a thin film composed of long-chain organic molecules with a head group that bonds directly to the surface and a functional tail self assembles into a monolayer in a dense brush-like configuration. SAMs were first discovered in the 1940s (Bigelow et al. 1946), but the field of surface science using SAMs developed rapidly from the 1980s onwards with the crucial development of thiol-SAMs where a sulphur head group binds to gold surfaces (Nuzzo and Allara 1983) and trichlorosilane-SAM using silane's affinity to silica (Maoz and Sagiv 1984). SAMs offer tight control over the surface characteristics. For example, a terminal carboxylic group will give a hydrophilic SAM surface, whereas an alkane tail group will give a hydrophobic SAM surface. This level of control and flexibility of SAM surface chemistry, along with the advent of nanoscale analytical instrumentation over this time, leads to a revolution in surface science over the last few decades led by researchers such as Whitesides (Qin et al. 2010). Generally, the physisorption of a hydrophobic protein or fraction of a protein onto a hydrophobic SAM can be used as a non-specific form of patterning, demonstrated in later sections (Arakaki et al. 2009; Martinez et al. 2011). For a more specific attachment, polyethylene-terminated alkanethiol (PEG) SAM molecules are used. These are hydrophobic and create a SAM surface on gold that resists biofouling by the protein. These can be patterned with hydrophilic carboxylic acid-terminated PEG-alkanethiols (PEG-COOH) where the COOH group can be activated to form a peptide bond with a primary amine of

the protein to attach it (Lahiri et al. 1999; Rusmini et al. 2007). This creates a surface with distinct regions of protein-attached and protein-repellent areas. However, this multistep process is not trivial, and more simple routes have been explored using the direct attachment of a protein to the substrate by modifying the protein with a cysteine amino acid or "silica affinity peptide" end group to strongly and specifically bind directly to the gold or silica surface, respectively. In this method a PEG SAM surface is patterned by particle attachment to an alternative SAM on the surface in the chosen pattern or removing some PEG leaving exposed a pattern of underlying substrate so the protein can bind directly to it (Bird et al. 2015, 2016a; b; Galloway et al. 2015, 2016).

3. *The method of patterning the surface*. In this section we discuss two patterning methods. The first is the soft lithographic technique of micro-contact printing (μCP)/nanocontact printing (nCP) which is a cheap and versatile stamping method. Here a PDMS elastomer stamp is "inked" with the chosen SAM molecule and printed onto the substrate surface. Once the master stamp is made, the method is easy to use and can accommodate a range of patterned designs. However, as it is a manual process, there are issues with reproducibility and quality of the pattern, limiting the patterning feature size to about 1 μm (below this more complex PDMS materials are required). To pattern reliably below this size, we consider a second patterning method: interferometric lithography (IL). IL uses the interference pattern of a split laser beam (using a Lloyds mirror arrangement) to etch a pattern in the SAM down to 20 nm scale features (Fig. 5.3).

The literature is summarised by these categories in table below (Table 5.2).

Table 5.2 Summary of protein and peptide directed magnetic nanoparticle arrays

		Patterning method		
		Micro-contact printing		IL
Material	Attachment to protein	SAM	Substrate	Substrate
Fe_3O_4	Mms6	Au-SAM (Galloway et al. 2012a, b)	Au-Cys (Bird et al. 2015, 2016b)	Au-Cys (Bird et al. 2016a)
		SiO_2-SAM (Arakaki et al. 2009)		
	Ferritin	SiO_2-SAM (Martinez et al. 2011)		
CoPt	Ferritin	No patterning, self-assembled on surface (Mayes et al. 2003)		
	Peptide		SiO_2-DAP (Galloway et al. 2015)	Au-Cys (Galloway et al. 2016)
			Au-Cys (Galloway et al. 2016)	

5.5.1 Mms6-Mediated Magnetite Patterned Surfaces

The first report of protein-mediated synthesis of magnetite on a surface was in 2009 where Mms6 was used to direct the location of magnetite MNP synthesis (Arakaki et al. 2009). An octadecyltrimethoxysilane (OTS) SAM molecule was used to form a hydrophobic SAM on a silica surface. A macroscale pattern was formed by simply pipetting Mms6 onto the surface. Mms6 is thought to physisorb to the hydrophobic surface via its N-terminal region, presumably exposing the more hydrophilic iron ion binding region to the mineralisation process. Contact angle measurements showed the hydrophobic surface became hydrophilic after protein addition. Iron oxide was mineralised on the Mms6 areas, but not on the hydrophobic areas of the surface. Mineralisation was not seen on areas covered in a control protein, BSA (Arakaki et al. 2009).

In 2012 this idea was advanced by Galloway et al. who improved this protocol by attaching Mms6 to the surface specifically by the N-terminal amine, ensuring the protein is in the correct orientation (with the C-termini active and exposed) to control magnetite MNP synthesis (Galloway et al. 2012a). This method was the first to demonstrate μm patterning using a μCP method to form 20 μm squares. Here a hydrophobic PEG SAM was patterned on a gold surface, and a 10% PEG-COOH SAM was backfilled into the non-SAM patterned regions. Proteins can be bound to COOH groups via their N-termini through a covalent peptide bond.

Ethyl(dimethylaminopropyl) carbodiimide (EDC) and N-hydroxysuccinimide (NHS) form an activated ester on a carboxylic acid surface, which preferentially bind to the N-terminal amine of a protein, when the protein is buffered to ≈1 pH point below its isoelectric point (pI) (Lahiri et al. 1999; Rusmini et al. 2007). This protocol was subjected to extensive optimisation for different mineralisation conditions and analysis (Galloway et al. 2012a, b) and showed that the magnetite MNPs formed on the surface have a very narrow size distribution but are much larger than MNPs formed in an Mms6-mediated in vitro synthesis and bigger than those grown in the native magnetosome. Although this method clearly demonstrates the activity of Mms6 and its ability to pattern MNP on surfaces, the EDC/NHS attachment adds an additional synthetic step. Therefore the attachment method was altered by modifying the Mms6 protein to bind directly to the substrate. A cysteine residue was added to the N-terminal region of Mms6. The sulphur atom of the cysteine binds directly to the gold surface. The SAM patterning protocol was unchanged, but now instead of backfilling with a PEG-COOH SAM to bind the protein, the Cys-Mms6 protein backfills directly onto the exposed gold areas. This was successfully achieved by Bird et al. in 2015, and importantly the patterning was much cleaner (Fig. 5.3f-h) (Bird et al. 2015). Furthermore, this meant that the size of the pattern could be reduced to the μCP limit of 1 μm wide lines that showed excellent definition between the MNPs and the SAM (Bird et al. 2015). Importantly the MNPs formed on these surfaces were smaller than the SAM bound Mms6 surface MNP. At 90 ± 15 nm they are closer in size to native magnetosomes and have tighter population distribution.

The patterning methods so far have gone from the macro- to the microscale, but are not yet on the nanoscale. Therefore, the cysteine-modified Mms6 protocol was advanced in 2016 by Bird et al. with the patterning method of IL (Fig. 5.3a-c) (Bird et al. 2016a). A PEG SAM surface was prepared and 350 nm wide lines were etched to expose the underlying gold, and the Cys-Mms6 protein was deposited into these areas. Mineralisation on these surfaces occurs with high pattern definition and yields well-defined magnetite MNPs of a size and distribution in agreement with the previous study (86 ± 21 nm). The size increase of MNP on surfaces compared to the solution phase and the native magnetosome may be a result of the curvature on the environment the Mms6 protein is displayed in (see Sect. 5.2) (Bird et al. 2016b). Work with Mms6 demonstrates the potential of this methodology to pattern arrays on MNPs on surfaces down to the nanoscale for potential nanodevice fabrication. However, although magnetite has many uses, it is not the ideal material for data storage, and Mms6 restricts the size of the MNPs to dimensions which are too large for many applications.

5.5.2 *Ferritin-Mediated MNP Patterned Surfaces*

Ferritin has been successfully electrostatically absorbed on μCP patterned surfaces, reported by Martinez et al. in 2011 (Martinez et al. 2011). Approximately 360 nm wide OTS SAMs were stamped on a silica surface, and this was backfilled with 3-aminopropyltriethoxysilane (APTES). The negatively charged ferritin adsorbed onto the positively charged amino-terminated APTES, giving strips approximately 400 nm wide. Interestingly, by varying the pressure on the stamp, only the edges of the strips can be printed leading to a different pattern of alternating 400 nm and 90 nm ferritin strips. The iron mineral inside the ferritin is ~7 nm ferrihydrite and thus poorly magnetic. In this study they heat their surfaces to 500 °C in O_2 to remove the ferritin protein cage. Further plasma etching in O_2 causes the size of the iron oxide particles to decrease to 2 nm. While this demonstrates good patterning of biotemplated iron oxides on surfaces within a ferritin cage, the heating step should perhaps be avoided as the effect this has on the iron oxide core has not been assessed and the magnetism of the particle cores are low. Interestingly, the ferritin cage has been used as a template to mineralise other MNPs (see Sect. 5.3). However, we have only found one example of these immobilised on a surface reported by Mayes et al. in 2003 (Mayes et al. 2003). CoPt was deposited within apoferritin to form 7 nm CoPt particles. These were predominately the semicrystalline fcc disordered phase. They were assembled on a silica substrate surface which was then subjected to high-temperature (500–650 °C) annealing to obtain the $L1_0$ phase. Mayes argues that the ferritin not only defines the MNPs size but also provides a shell to prevent sintering and agglomeration during the annealing step. Although the ferritin protein is removed during annealing, and the particles are seen to cluster somewhat after this step. This highlights the clear advantage in being able to remove all high-temperature processes related to fabrication. Most interestingly, the authors performed a contact drag test to assess information recording and reading. Although the surfaces need optimisation, there is a promise to this method as a clear read-back signal is demonstrated.

5.5.3 Peptide-Mediated CoPt Patterned Surfaces

5.5.3.1 Dual Affinity Peptide

Previous biopanning experiments carried out by Klem et al. in 2005 identified one key peptide sequence (KTHEIHSPLLHK) that binds to the $L1_0$ phase of CoPt (Klem et al. 2005). This CoPt-binding peptide has been used to create a dual affinity peptide (DAP) that is able to bind to a silica surface (Galloway et al. 2015). The DAP contains a peptide sequence known to bind to silica (HPPMNASHPHMH) connected to the CoPt-binding peptide by a flexible linker (GSG) so the DAP has the sequence HPPMNASHPHMH-GSG-KTHEIHSPLLHK (Eteshola et al. 2005; Klem et al. 2005). In this study surfaces were prepared via μCP. A PDMS stamp was inked with 1 mg/ml CoPt-DAP solution, and after 1 min excess solution was removed and the stamp dried. The inking and drying procedure was repeated at least ten times before the stamp was placed into conformal contact with a silicon surface. The surface was then placed into a reaction of Co(II) and Pt(II) salts under reducing conditions, at room temperature. This led to the formation of CoPt NPs attached to a silica surface via the DAP. X-ray diffraction data suggests that the DAP is able to biotemplate the $L1_0$ phase of CoPt (Galloway et al. 2015). This is due to the presence of (001) and (100) planes that are expected for CoPt $L1_0$ reflections. It is thought that the DAP is able to lower the activation energy for the formation of $L1_0$ phase in the aqueous phase. This is because the DAP is more closely packed when on a silica surface and therefore may enhance crystallographic alignment and improve the templating abilities of the peptide. Another possibility is that the peptide may organise particles into the tetragonal phase and aid in the reduction to metallic CoPt when bound to the surface (Galloway et al. 2015).

However, magnetisation data shows that the coercivity is very low and this could be because the biotemplated $L1_0$ MNPs are not aligned in the same way as sputtered $L1_0$ films. When hysteresis measurements have been carried out for sputtered films that have misaligned grains loops with reduced or no coercivity were observed. Therefore, it is possible that the magnetic orientations of the MNPs are not aligned correctly to achieve maximum coercivity.

5.5.3.2 Cys-CoPt Peptide

The peptide (KTHEIHSPLLHK) previously identified from biopanning as being specific to the $L1_0$ phase of CoPt has been further modified for attachment onto surfaces. The modification involved incorporation of a cysteine amino acid residue connected to the original peptide via a flexible linker (GSG), C-GSG-KTHEIHSPLLHK (Galloway et al. 2016).

In this study IL and μCP were used to pattern gold surfaces allowing nano- and microscale lines to be produced. In IL PEG SAMs were first formed on the surface and then exposed to the interferometer. Regions that were exposed to the interferometer resulted in removal of the SAM to reveal bare gold. For μCP the PEG SAM was applied to the surface using a PDMS stamp, and bare regions on both IL and μCP surfaces were then backfilled with cys-CoPt peptide before mineralisation of CoPt. SEM images of the biotemplated surfaces show that regions functionalised with cys-CoPt were covered in a densely packed thin film of MNPs, whereas PEG SAM regions remained free of any NPs. An average line width of 226 nm and 6.1 μm was observed for lines of CoPt NPs generated by IL (Fig. 5.3d-e) and μCP (Fig. 5.3i,j), respectively (Galloway et al. 2016). Grain size analysis was performed on these SEM images, and NPs mineralised using cys-CoPt had an average diameter of 12 nm (Galloway et al. 2016). A previous study that uses a dual affinity peptide to mineralise NPs had an average diameter of 17 nm (Galloway et al. 2015). Therefore this suggest that immobilisation of the cys-CoPt peptide onto a substrate enhances its ability to control the formation of uniform MNPs. As the cys-CoPt is shorter than the DAP, this could result in closer packing of the peptide onto the surface and enhance templating of smaller more uniform particles. EDX data for the biotemplated surfaces shows that CoPt has been formed in a 1:1 stoichiometric ratio by using the cys-CoPt peptide. This is typical for $L1_0$ CoPt. From XRD and SAED data, peaks that correspond to the (001), (101) and (110) planes of $L1_0$ CoPt were observed for the biotemplated surfaces. $CoPt_3$ peaks were also present, but values for these were not as close to literature values. Therefore, due to the EDX data, the peaks on the XRD were tentatively assigned as $L1_0$ CoPt. No $L1_0$ structure was observed for bulk particles, therefore suggesting that immobilisation of the cys-CoPt peptide to a surface shows control over the particles formed (Galloway et al. 2016).

Magnetic data indicated that the nanoparticles formed on the surface were ferromagnetic at room temperature with coercivities of 20–30 Oe. This is lower than expected for $L1_0$ CoPt. To achieve high coercivity of the $L1_0$ phase, the c-axis needs to be aligned perpendicular to the surface (Galloway et al. 2016). Therefore, as the biotemplated surfaces in this study have a low coercivity, it suggests that they lack this c-axis alignment. Mineralisation was also carried out in the presence of a 0.2 T magnetic field; these surfaces did not show any increase in coercivity and had a lack of long-range ordering.

5.6 Conclusion

This chapter has shown that specific proteins can act as powerful tools in controlling the synthesis of magnetic nanoparticles. Proteins derived from the magnetosome, such as Mms6, MmsF and MamC, display remarkable self-assembly properties when studied in vitro and retain the ability to reduce the heterogeneity of nanoparticle products in synthetic MNP precipitation reactions. Not only can these proteins be used synthetically, but they provide a valuable model system for understanding protein-mediated biomineralisation. They also provide a proof of concept for the successful utilisation of proteins in magnetic material synthesis, and the field has now widened to encompass designed proteins such as the MIA and the re-engineering of naturally occurring proteins such as the heat shock protein MjHsp.

One important element of the magnetosome, which is often overlooked, is the presence of the lipid vesicle. Constraining the synthesis of magnetite within the lumen of the magnetosome offers a level of size control of the magnetite nanoparticle which is hard to achieve any other way. Studying biomineralisation proteins such as Mms6 in a water-soluble micelle may be quite different compared to when the protein is embedded in the magnetosome membrane. An important future direction for this field will be the incorporation of such proteins into vesicle structures to create artificial magnetosomes. This approach would provide two distinct advantages over solution phase synthesis: (i) it offers a close to native environment in which to study the mechanism of each protein, and (ii) it can provide a further level of control over MNP synthesis and may result in particles with a higher degree of homogeneity than can be achieved with the protein alone. Steps towards this goal have already been made, with the successful precipitation of iron oxides within lipid and polymer vesicles (Bain et al. 2015a, b; Bakhshi et al. 2016). Introducing proteins into this type of system could lead to new advances and insights.

In addition to the benefits on particle formation, if used on surfaces, these proteins have a dual role of forming and binding the particles while simultaneously anchoring them to the surface. This has been demonstrated successfully with Mms6 on gold surfaces patterned with both IL and micro-contact printing. A similar approach has been used with CoPt nanoparticles using a dual affinity peptide which has binding interactions with both the CoPt NP and the silica surface. These approaches demonstrate the flexibility available to the protein engineer, as a single protein can be designed to fulfil two key roles.

We are entering a critical era for protein-mediated nanomaterial synthesis. The understanding gleaned from the wealth of previous studies, some of which are highlighted in this chapter, coupled with advances in material synthesis, protein engineering and new patterning technologies, will lead to new innovations in this fast-moving field. A future with cleaner, cheaper and more precise magnetic nanoparticles is moving a step closer to reality.

References

Aboaf J, Herd S, Klokholm E (1983) Magnetic-properties and structure of cobalt platinum thin-films. IEEE Trans Magn 19:1514–1519. https://doi.org/10.1109/TMAG.1983.1062575

Aljabali AAA, Barclay JE, Cespedes O, Rashid A, Staniland SS, Lomonossoff GP, Evans DJ (2011a) Charge modified cowpea mosaic virus particles for templated mineralization. Adv Funct Mater 21:4137–4142

Aljabali AAA, Shah SN, Evans-Gowing R, Lomonossoff GP, Evans DJ (2011b) Chemically-coupled-peptide-promoted virus nanoparticle templated mineralization. Integr Biol 3:119–125

Allen M, Willits D, Mosolf J, Young M, Douglas T (2002) Protein cage constrained synthesis of ferrimagnetic iron oxide nanoparticles. Adv Funct Mater 14:1562–1565

Allen M, Willits D, Young M, Douglas T (2003) Constrained synthesis of cobalt oxide nanomaterials in the 12-subunit protein cage from Listeria innocua. Inorg Chem 42:6300–6305

Amemiya Y, Arakaki A, Staniland SS, Tanaka T, Matsunaga T (2007) Controlled formation of magnetite crystal by partial oxidation of ferrous hydroxide in the presence of recombinant magnetotactic bacterial protein Mms6. Biomaterials 28:5381–5389. https://doi.org/10.1016/j.biomaterials.2007.07.051

Arakaki A, Webb J, Matsunaga T (2003) A novel protein tightly bound to bacterial magnetic particles in Magnetospirillum magneticum strain AMB-1. J Biol Chem 278:8745–8750. https://doi.org/10.1074/jbc.M211729200

Arakaki A, Masuda F, Matsunaga T (2009) Iron oxide crystal formation on a substrate modified with the Mms6 protein from magnetotactic bacteria. MRS Online Proc Libr Arch 1187:1187–KK1103-1108. (1185 pages. https://doi.org/10.1557/PROC-1187-KK03-08

Arakaki A, Masuda F, Amemiya Y, Tanaka T, Matsunaga T (2010) Control of the morphology and size of magnetite particles with peptides mimicking the Mms6 protein from magnetotactic bacteria. J Colloid Interface Sci 343:65–70. S0021–9797(09)01488-X [pii]. https://doi.org/10.1016/j.jcis.2009.11.043

Bain J, Berry M, Dirks C, Staniland S (2015a) Synthesis of ABA tri-block Co-polymer magnetopolymersomes via electroporation for potential medical application. Polymer 7:2558–2571. https://doi.org/10.3390/polym7121529

Bain J, Ruiz-Perez L, Kennerley A, Muench S, Thompson R, Battaglia G, Staniland S (2015b) In situ formation of magnetopolymersomes via electroporation for MRI. Sci Rep 5. https://doi.org/10.1038/srep14311

Bakhshi PK, Bain J, Gul MO, Stride E, Edirisinghe M, Staniland SS (2016) Manufacturing man-made Magnetosomes: high-throughput in situ synthesis of biomimetic magnetite loaded Nanovesicles. Macromol Biosci. https://doi.org/10.1002/mabi.201600181

Baumgartner J, Carillo MA, Eckes KM, Werner P, Faivre D (2014) Biomimetic magnetite formation: from biocombinatorial approaches to mineralization effects. Langmuir 30:2129–2136. https://doi.org/10.1021/la404290c

Bigelow WC, Pickett DL, Zisman WA (1946) Oleophobic monolayers. J Colloid Sci 1:513–538. https://doi.org/10.1016/0095-8522(46)90059-1

Bird SM, Galloway JM, Rawlings AE, Bramble JP, Staniland SS (2015) Taking a hard line with biotemplating: cobalt-doped magnetite magnetic nanoparticle arrays. Nanoscale 7:7340–7351. https://doi.org/10.1039/C5NR00651A

Bird SM, El-Zubir O, Rawlings AE, Leggett GJ, Staniland SS (2016a) A novel design strategy for nanoparticles on nanopatterns: interferometric lithographic patterning of Mms6 biotemplated magnetic nanoparticles. J Mater Chem C. https://doi.org/10.1039/C5TC03895B

Bird SM, Rawlings AE, Galloway JM, Staniland SS (2016b) Using a biomimetic membrane surface experiment to investigate the activity of the magnetite biomineralisation protein Mms6. RSC Adv 6:7356–7363. https://doi.org/10.1039/C5RA16469A

Bozzi M, Mignogna G, Stefanini S, Barra D, Longhi C, Valenti P, Chiancone E (1997) A novel non-heme Iron-binding ferritin related to the DNA binding proteins of the Dps family in Listeria innocua J Biol Chem 272:3259–3265

Burkhard P, Stetefeld J, Strelkov S (2001) Coiled coils: a highly versatile protein folding motif. Trends Cell Biol 11:82–88. https://doi.org/10.1016/S0962-8924(00)01898-5

Charap SH, Lu P-L, He Y (1997) Thermal stability of recorded information at high densities magnetics. IEEE Trans on 33:978–983

del Puerto MM, Gonzalez-Carreno T, Tartaj P, Veintemillas-Verdaguer S, Serna CJ (2003) The preparation of magnetic nanoparticles for applications in biomedicine. J Phys D Appl Phys 36:R182–R197. https://doi.org/10.1088/0022-3727/36/13/202

Douglas T, Strable E, Willits D, Aitouchen A, Libera M, Young M (2002) Protein engineering of a viral cage for constrained nanomaterials synthesis. Adv Mater 14:415–418

Driskill-Smith AA (2004) Electron-beam and emerging lithography for the magnetic recording industry. In: Microlithography 2004. International Society for Optics and Photonics, pp 16–30

Eteshola E, Brillson LJ, Lee SC (2005) Selection and characteristics of peptides that bind thermally grown silicon dioxide film. Biomol Eng 22:201–204

Feng SR, Wang LJ, Palo P, Liu XP, Mallapragada SK, Nilsen-Hamilton M (2013) Integrated self-assembly of the Mms6 magnetosome protein to form an Iron-responsive structure. Int J Mol Sci 14:14594–14606. https://doi.org/10.3390/ijms140714594

Flenniken ML, Willits DA, Brumfield S, Young MJ, Douglas T (2003) The small heat shock protein cage from Methanococcus jannaschii is a versatile nanoscale platform for genetic and chemical modification ACS. Nano Lett 3:1573–1576

Galloway JM, Arakaki A, Masuda F, Tanaka T, Matsunaga T, Staniland SS (2011) Magnetic bacterial protein Mms6 controls morphology, crystallinity and magnetism of cobalt-doped magnetite nanoparticles in vitro. J Mater Chem 21:15244–15254. https://doi.org/10.1039/C1jm12003d

Galloway JM, Bramble JP, Rawlings AE, Burnell G, Evans SD, Staniland SS (2012a) Biotemplated magnetic nanoparticle arrays. Small 8:204–208. https://doi.org/10.1002/smll.201101627

Galloway JM, Bramble JP, Rawlings AE, Burnell G, Evans SD, Staniland SS (2012b) Nanomagnetic arrays formed with the biomineralization protein Mms6. J Nano Res Sw 17:127–146. https://doi.org/10.4028/www.scientific.net/JNanoR.17.127

Galloway JM, Talbot JE, Critchley K, Miles JJ, Bramble JP (2015) Developing biotemplated data storage: room temperature biomineralization of L10 CoPt magnetic nanoparticles. Adv Funct Mater 25:4590–4600. https://doi.org/10.1002/adfm.201501090

Galloway JM et al (2016) Nano- and micro-patterning biotemplated magnetic CoPt arrays. Nanoscale 8:11738–11747. https://doi.org/10.1039/C6NR03330J

Gurnon DG, Whitaker J, Oakley MG (2003) Design and characterization of a homodimeric antiparallel coiled coil. J Am Chem Soc 125:7518–7519. https://doi.org/10.1021/ja0357590

Hatch GP, Stelter RE (2001) Magnetic design considerations for devices and particles used for biological high-gradient magnetic separation (HGMS) systems. J Magn Magn Mater 225:262–276. https://doi.org/10.1016/S0304-8853(00)01250-6

Huang J, Bu L, Xie J, Chen K, Cheng Z, Li X, Chen X (2010) Effects of nanoparticle size on cellular uptake and liver MRI with polyvinylpyrrolidone-coated iron oxide nanoparticles. ACS Nano 4:7151–7160. https://doi.org/10.1021/nn101643u

Jijakli K et al (2016) The in vitro selection world. Methods 106:3–13. https://doi.org/10.1016/j.ymeth.2016.06.003

Jutz G, van Rijn P, Miranda BS, Boker A (2015) Ferritin: A Versatile Building Block for Bionanotechnology. Chem Rev 115:1653–1701. https://doi.org/10.1021/cr400011b

Kashyap S, Woehl TJ, Liu X, Mallapragada SK, Prozorov T (2014) Nucleation of iron oxide nanoparticles mediated by Mms6 protein in situ. ACS Nano 8:9097–9106. https://doi.org/10.1021/nn502551y

Kim KK, Kim R, Kim S-H (1998) Crystal structure of a small heat-shock protein. Nature 394:595–599

Klem MT, Willits D, Solis DJ, Belcher AM, Young M, Douglas T (2005) Bio-inspired synthesis of protein encapsulated CoPt nanoparticles. Adv Funct Mater 15:1489–1494

Lahiri J, Isaacs L, Tien J, Whitesides GM (1999) A strategy for the generation of surfaces presenting ligands for studies of binding based on an active ester as a common reactive intermediate: a surface plasmon resonance study. Anal Chem 71:777–790

Laughlin D, Srinivasan K, Tanase M, Wang L (2005) Crystallographic aspects of L1(0) magnetic materials. Scr Mater 53:383–388. https://doi.org/10.1016/j.scriptamat.2005.04.039

Laurent S, Forge D, Port M, Roch A, Robic C, Vander Elst L, Muller RN (2008) Magnetic iron oxide nanoparticles: synthesis, stabilization, vectorization, physicochemical characterizations, and biological applications. Chem Rev 108:2064–2110. https://doi.org/10.1021/cr068445e

Lee N et al (2011) Magnetosome-like ferrimagnetic iron oxide nanocubes for highly sensitive MRI of single cells and transplanted pancreatic islets. Proc Natl Acad Sci U S A 108:2662–2667. https://doi.org/10.1073/pnas.1016409108

Love AJ, Makarov V, Yaminsky I, Kalinina NO, Taliansky ME (2014) The use of tobacco mosaic virus and cowpea mosaic virus for the production of novel metal nanomaterials. Virology 449:133–139

Lu AH, Salabas EL, Schuth F (2007) Magnetic nanoparticles: synthesis, protection, functionalization, and application. Angew Chem Int Ed 46:1222–1244. https://doi.org/10.1002/anie.200602866

Maoz R, Sagiv J (1984) On the formation and structure of self-assembling monolayers. I. A comparative atr-wettability study of Langmuir—Blodgett and adsorbed films on flat substrates and glass microbeads. J Colloid Interface Sci 100:465–496. https://doi.org/10.1016/0021-9797(84)90452-1

Martın J, Nogues J, Liu K, Vicent J, Schuller IK (2003) Ordered magnetic nanostructures: fabrication and properties. J Magn Magn Mater 256:449–501

Martinez RV, Chiesa M, Garcia R (2011) Nanopatterning of ferritin molecules and the controlled size reduction of their magnetic cores. Small 7:2914–2920. https://doi.org/10.1002/smll.201100366

Mayes E et al (2003) Biologically derived nanomagnets in self-organized patterned media. IEEE Trans Magn 39:624–627. https://doi.org/10.1109/tmag.2003.808982

Meldrum FC, Wade VJ, Nimmo DL, Heywood BR, Mann S (1991) Synthesis of inorganic nanophase materials in supramolecular protein cages. Nature 349:684–687

Mikata Y, He Q, Lippard SJ (2001) Laser-induced photo-cross-linking of cisplatin-modified DNA to HMG-domain proteins. Biochemistry 40:7533–7541

Moore LR et al (2001) Progenitor cell isolation with a high- capacity quadrupole magnetic flow sorter. J Magn Magn Mater 225:277–284. https://doi.org/10.1016/S0304-8853(00)01251-8

Moroz P, Jones SK, Gray BN (2001) Status of hyperthermia in the treatment of advanced liver cancer. J Surg Oncol 77:259–269. https://doi.org/10.1002/jso.1106

Murat D et al (2012) The magnetosome membrane protein, MmsF, is a major regulator of magnetite biomineralization in Magnetospirillum magneticum AMB-1. Mol Microbiol 85:684–699. https://doi.org/10.1111/j.1365-2958.2012.08132.x

Muthana M et al (2015) Directing cell therapy to anatomic target sites in vivo with magnetic resonance targeting. Nat Commun:6. https://doi.org/10.1038/ncomms9009

Nickolls J, Dally WJ (2010) The GPU computing era. IEEE Micro 30:56–69

Nudelman H et al (2016) Structure-function studies of the magnetite-biomineralizing magnetosome-associated protein MamC. J Struct Biol 194:244–252. https://doi.org/10.1016/j.jsb.2016.03.001

Nuzzo RG, Allara DL (1983) Adsorption of bifunctional organic disulfides on gold surfaces. J Am Chem Soc 105:4481–4483. https://doi.org/10.1021/ja00351a063

Pankhurst QA, Connolly J, Jones SK, Dobson J (2003) Applications of magnetic nanoparticles in biomedicine. J Phys D Appl Phys 36:R167–R181. https://doi.org/10.1088/0022-3727/36/13/201

Paul F, Melville D, Roath S, Warhurst D (1981) A bench top magnetic separator for malarial parasite concentration magnetics. IEEE Trans on 17:2822–2824. https://doi.org/10.1109/TMAG.1981.1061711

Pontillo N, Pane F, Messori L, Amoresano A, Merlino A (2016) Cisplatin encapsulation within a ferritin nanocage: a high-resolution crystallographic study. Chem Commun 52:4136–4139. https://doi.org/10.1039/c5cc10365g

Prozorov T et al (2007) Protein-mediated synthesis of uniform superparamagnetic magnetite nanocrystals. Adv Funct Mater 17:951–957. https://doi.org/10.1002/adfm.200600448

Qin D, Xia Y, Whitesides GM (2010) Soft lithography for micro- and nanoscale patterning. Nat Protoc 5:491–502

Rawlings A (2016) Membrane proteins: always an insoluble problem? Biochem Soc Trans 44:790–795. https://doi.org/10.1042/BST20160025

Rawlings AE, Bramble JP, Walker R, Bain J, Galloway JM, Staniland SS (2014) Self-assembled MmsF proteinosomes control magnetite nanoparticle formation in vitro. Proc Natl Acad Sci USA 111:19094–19099. https://doi.org/10.1073/pnas.1409256111

Rawlings AE et al (2015) Phage display selected magnetite interacting Adhirons for shape controlled nanoparticle synthesis. Chem Sci. https://doi.org/10.1039/C5SC01472G

Rawlings AE, Bramble JP, Hounslow AM, Williamson MP, Monnington AE, Cooke DJ, Staniland SS (2016) Ferrous iron key to Mms6 magnetite biomineralisation:a mechanistic study to understand magnetite formation using pH titration and NMR. Chem Eur J 22:7885–7894

Regazzoni AE, Urrutia GA, Blesa MA, AJG M (1981) Some observations on the composition and morphology of synthetic magnetites obtained by different routes. J Inorg Nucl Chem 43:1489–1493

Reiss BD, Mao C, Solis DJ, Ryan KS, Thomson T, Belcher AM (2004) Biological routes to metal alloy ferromagnetic nanostructures ACS. Nano Lett 4:1127–1132

Richter HJ et al (2006) Recording on bit-patterned media at densities of 1 Tb/in <img src="/images/tex/413.Gif" alt="^2"> and beyond. IEEE Trans Magn 42:2255–2260. https://doi.org/10.1109/TMAG.2006.878392

Rusmini F, Zhong Z, Feijen J (2007) Protein immobilization strategies for protein biochips. Biomacromolecules 8:1775–1789. https://doi.org/10.1021/bm061197b

San BH, Lee S, Moh SH, Park J-G, Lee JH, Hwang H-Y, Kim KK (2013) Size-controlled synthesis and characterization of CoPt nanoparticles using protein shells. J Mater Chem B t1:1453. https://doi.org/10.1039/c2tb00290f

Sanchez P et al (2009) MRI relaxation properties of water-soluble apoferritin-encapsulated gadolinium oxide-hydroxide nanoparticles. Dalton Trans:800–804. https://doi.org/10.1039/b809645g

Sangregorio C, Wiemann J, Connor CJ, Rosenzweig Z (1999) A new method for the synthesis of magnetoliposomes. J Appl Phys 85:5699–5701

Sano K, Ajima K, Iwahori K, Yudasaka M, Iijima S, Yamashita I, Shiba K (2005) Endowing a ferritin-like cage protein with high affinity and selectivity for certain inorganic materials. Small 1:826–832. https://doi.org/10.1002/smll.200500010

Sengonul M, Ruzicka J, Attygalle AB, Libera M (2007) Surface modification of protein nanocontainers and their self-directing character in polymer blends. Polymer 48:3632–3640. https://doi.org/10.1016/j.polymer.2007.04.017

Smith GP (1985) Filamentous fusion phage: novel expression vectors that display cloned antigens on the virion surface. Science 228:1315–1317

Staniland S, Rawlings A (2016) Crystallizing the function of the magnetosome membrane mineralization protein Mms6. Biochem Soc Trans 44:883–890. https://doi.org/10.1042/BST20160057

Sun S et al (2003) Controlled synthesis and assembly of FePt nanoparticles. J Phys Chem B 107:5419–5425

Sun X, Jia Z, Huang Y, Harrell J, Nikles D, Sun K, Wang L (2004) Synthesis and magnetic properties of CoPt nanoparticles. J Appl Phys 95:6747–6749. https://doi.org/10.1063/1.1667441

Tanaka M, Mazuyama E, Arakaki A, Matsunaga T (2010) Mms6 protein regulates crystal morphology during nano-sized magnetite biomineralization in vivo. J Biol Chem. M110.183434 [pii]. https://doi.org/10.1074/jbc.M110.183434

Terris B, Thomson T (2005) Nanofabricated and self-assembled magnetic structures as data storage media. J Phys D Appl Phys 38:R199

Terris B, Thomson T, Hu G (2007) Patterned media for future magnetic data storage. Microsyst Technol 13:189–196

Tiede C et al (2014) Adhiron: a stable and versatile peptide display scaffold for molecular recognition applications. Protein Eng Des Sel 27:145–155. https://doi.org/10.1093/protein/gzu007

Valverde-Tercedor C et al (2015) Size control of in vitro synthesized magnetite crystals by the MamC protein of Magnetococcus marinus strain MC-1. Appl Microbiol Biotechnol 99:5109–5121. https://doi.org/10.1007/s00253-014-6326-y

van der Zee J (2002) Heating the patient: a promising approach? Ann Oncol 13:1173–1184. https://doi.org/10.1093/annonc/mdf280
Wang LJ et al (2012) Self-assembly and biphasic Iron-binding characteristics of Mms6, a bacterial protein that promotes the formation of superparamagnetic magnetite nanoparticles of uniform size and shape. Biomacromolecules 13:98–105. https://doi.org/10.1021/Bm201278u
Wang Y, Zhang X, Liu Y, Jiang Y, Zhang Y, Yang J (2014) Microstructure and magnetic properties of L1(0) CoPt nanoparticles by ag addition. J Sol-Gel Sci Technol 70:528–533. https://doi.org/10.1007/s10971-014-3318-4
Weller D et al (2000) High K-u materials approach to 100 Gbits/in(2). IEEE Trans Magn 36:10–15. https://doi.org/10.1109/20.824418
Wollhofen R, Katzmann J, Hrelescu C, Jacak J, Klar TA (2013) 120 nm resolution and 55 nm structure size in STED-lithography. Opt Express 21:10831–10840
Wong KKW, Colfen H, Whilton NT, Douglas T, Mann S (1999) Synthesis and characterization of hydrophobic ferritin proteins. J Inorg Biochem 76:187–195. https://doi.org/10.1016/s0162-0134(99)00114-2
Yamagishi A, Tanaka M, Lenders J, Thiesbrummel J, Sommerdijk N, Matsunaga T, Arakaki A (2016) Control of magnetite nanocrystal morphology in magnetotactic bacteria by regulation of mms7 gene expression. Sci Rep 6. https://doi.org/10.1038/srep29785
Yoshino T, Matsunaga T (2006) Efficient and stable display of functional proteins on bacterial magnetic particles using Mms13 as a novel anchor molecule. Appl Environ Microbiol 72:465–471. https://doi.org/10.1128/AEM.72.1.465-471.2006
Zhang YJ et al (2011) A novel approach to the synthesis of CoPt magnetic nanoparticles. J Phys D Appl Phys 44:295003
Zhang H et al (2015) Morphological transformations in the magnetite Biomineralizing protein Mms6 in Iron solutions: a small-angle X-ray scattering study. Langmuir 31:2818–2825. https://doi.org/10.1021/la5044377
Zhou C-Z, Confalonieri F, Jacquet M, Perasso R, Li Z-G, Janin J (2001) Silk fibroin: structural implications of a remarkable amino acid sequence proteins: structure. Funct Bioinforma 44:119–122. https://doi.org/10.1002/prot.1078

Chapter 6
Fabrication of Nanodevices Using Nanoparticles Synthesized in Cage-Shaped Proteins

Ichiro Yamashita

Abstract Fabrication of nanostructures by biomolecules, termed the "bio nano process (BNP)," has been proposed, and the process utilizes the biotemplated biomineralization of inorganic materials and self-assembly. NP synthesis by genetically modified cage-shaped proteins, ferritin, and DNA-binding protein from starved cells, Dps, is the most basic part of the BNP. The BNP can produce homogenous NPs inside cage-shaped protein. NP growth is limited by the protein shell and the same size NP can be produced. The surrounding protein shells deliver/array the accommodated NPs at designated positions through the interaction of the protein and inorganic material surface. Even a single NP placement to a nanodisk on a substrate is possible. In such placement and arraying, aptamer and electrostatic interactions play the important roles. Obtained NP arrays are proved to be able to fabricate key components of nanoelectronic devices through the integration of top-down and bottom-up technology. The BNP can also produce three-dimensional bioconjugate of cage-shaped proteins and CNTs. This new material has a quantum effect and high thermal insulation by the protein shell. The measurement of thermoelectric properties confirmed that this bioconjugate is a new type of thermoelectric materials.

6.1 Introduction

Nanoparticle (NP) is one of the most fundamental nanotechnology components. The nanoparticles (NPs) have intriguing functions originated from their small sizes, such as quantum effects, high catalytic activities, superparamagnetism, and

I. Yamashita (✉)
Panasonic Science Research Alliance laboratories, Graduate School of Engineering, Osaka University, Suita, Japan

Graduate School of Materials Science, Nara Institute of Science and Technology (NAIST), Ikoma, Japan
e-mail: yamashita@pmdp.arl.eng.osaka-u.ac.jp

T. Matsunaga et al. (eds.), *Biological Magnetic Materials and Applications*,
https://doi.org/10.1007/978-981-10-8069-2_6

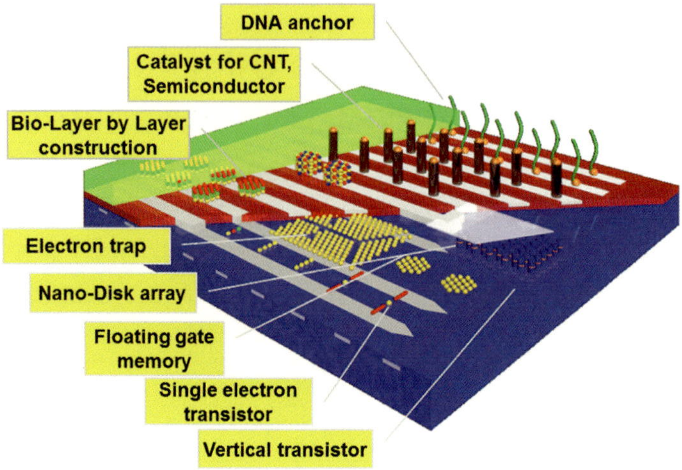

Fig. 6.1 Schematic drawing of the key components of nanoelectronic devices which the bio nano process is proposed to produce

enhanced surface plasmon, which cannot be realized by bulk materials. When the NPs are arrayed or constructed in two-dimensional/three-dimensional (2D/3D) structures, new functions emerge, which we cannot imagine by simply summing up NP's properties. Emerging new functions are deriving from the spatial distribution and nanometoric size itself. NPs have the potential for revolutionizing material design and producing completely new functional nanostructured material. These properties have fascinated researchers these several tens of years. A wide range of NP applications, such as magnetic recording media, biomedical magnetic beads, catalysts, fluorescent markers, sensors, ink, drug delivery systems, and quantum electronics, are being studied intensely.

Lots of methods have been studied, including physical, chemical, and biological methods for NP synthesis. Each method has its own specificities to satisfy application demands. Among those demands, homogenous size and dispersity in solution are commonly important. Since the NP properties are strongly connected to their sizes, nonuniformity directly leads to the fluctuation of their properties. NPs which are easy to aggregate and/or form collective clumps are cumbersome to handle, and such NP clumps lose their splendid properties. For the construction of 2D/3D functional nanostructures, homogenous and dispersive NPs are indispensable.

About 15 years ago, we proposed a biological process which produces nanoelectronic key components utilizing biomineralization and self-assembly of biomolecules, which was named bio nano process (BNP) (Yamashita 2000). Figure 6.1 shows schematic drawing of some examples of the functional nanostructures proposed to be produced by the BNP. The BNP utilizes a wide range of biotemplates to produce NP, nano-wire, nanotube, and other nanomaterials.

Biological NP synthesis is one of the most intensely studied fields of the BNP. NPs are synthesized in the cavity of cage-shaped protein (biotemplate). The methods have essential merits. Firstly, the cage-shaped proteins have atomically the same structure. Although many kinds of proteins alternate their tertiary structures depending on the circumstances, there are many structural proteins which have the same structure. Cage-shaped proteins are such examples including many kinds of virus protein shells. The X-ray crystallography technique clearly shows their shell structures in atomic resolution, which indicates their structure is atomically the same. When NPs are synthesized in the cavity, the size of the NPs should be the same, if the NP developed fully inside the cavity. Secondly, NPs are covered by the protein shell. The NPs are protected by the protein shell which makes the NP-protein bioconjugate dispersive. The protein surface is able to be modified genetically and control the interaction with other proteins, biomolecules, and inorganic materials. This provides tremendous merits. The bioconjugate can make 2D/3D structure spontaneously or in a self-assembly manner. The bioconjugates could be aligned on some area of the substrate by its specific interaction between protein shell and pattered layer on a substrate. Furthermore, bioconjugate can make 2D/3D nanostructures through their interactions. The fabrication of a functional nanostructure far smaller than the micro-objects is made much easier by using NPs produced with cage-shaped protein template.

6.2 Biotemplate for Nanoparticle Synthesis, Ferritin, and Dps

There are many cage-shaped proteins applicable for NP synthesis. The BNP has been focusing on ferritin and Dps (DNA-binding protein from starved cells). Ferritin is a ubiquitous iron storage protein among wide variety of lives (Fig. 6.2). Since excessive iron ions are toxic under oxidative conditions, the iron ions beyond the normal limits are stored in a ferritin as a ferrihydrite core, and the iron ions are supplied back as ferrous ion when required. The horse spleen apoferritin (HsAFr) structure was solved by X-ray crystallography, and the structure and function have been

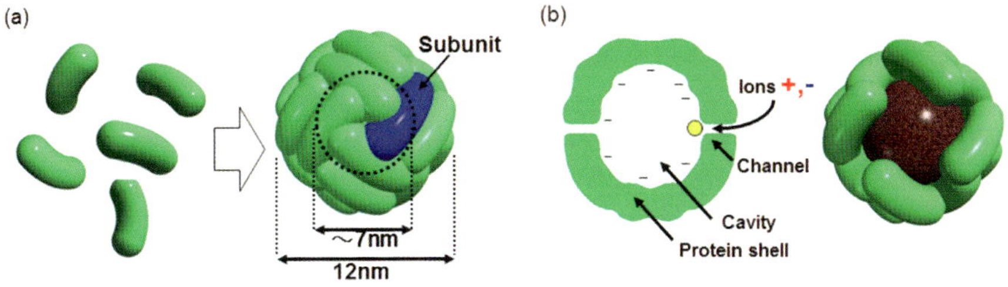

Fig. 6.2 Schematic drawing of ferritin assembly where 24 subunits self-assemble into a cage-shaped protein shell (**a**). Iron ions go through the threefold channel and are condensed around nucleation site to produce ferrihydrite core (**b**). A ferritin with NP core with four subunits is eliminated (**b**)

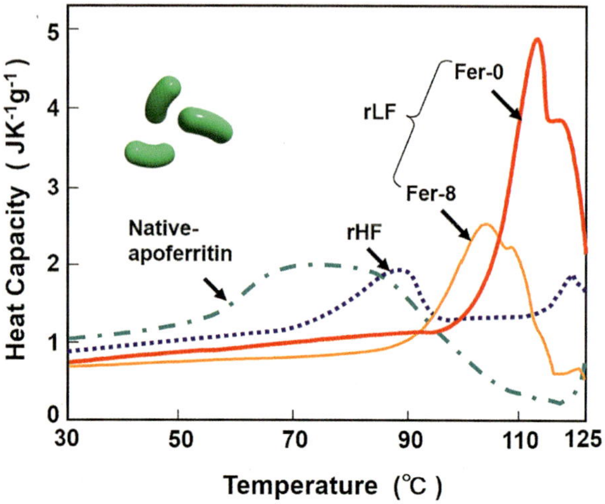

Fig. 6.3 Thermal stabilities of ferritins. Native ferritin is the horse spleen apoferritin. rHF and rLF stand for recombinant H- and L-apoferritins. Fer-8 lacks the first eight amino acid residues from N-terminus

well studied (Clegg et al. 1980; Levi et al. 1988; Harrison et al. 1991; Lawson et al. 1991; Massover 1993; Gallois et al. 1997). Other kinds of ferritins were also solved by X-ray crystallography, and their structures are surprisingly the same. The subunits are 4-alpha-helix bundles, and 24 subunits form a spherical cage-shaped protein. Its outer and inner diameters are 12 nm and 7 nm, respectively. Native protein supramolecule shell is composed of 24 heterogamous subunits and H- and L-chains. Their molecular weights are 21 kDa and 18 kDa, respectively, and the ratio of the H- and L-chains varies depending on species and tissues. In the case of horse spleen ferritin, the constituent ratio is 90% L-chains and 10% H-chain. Fifty-five percent of their sequences are identical, and they form a bundle of four antiparallel helices (A–D) and a C-terminal short helix (E) and a long loop connecting helices B and C. The H-chain has a metal-binding site which has ferroxidase activity, and L-chain has Cd-binding sites which can make salt bridge with other L-chains. The subunits are preferentially organized in dimeric building blocks and ideally suited to the generation of a molecular assembly, protein shell, with 432 symmetry. There are eight small inter-subunit channels connecting the outside and inner cavity along threefold axis (threefold channel) which is adopted as the passage of iron ions. The cage-shaped protein without core is called apoferritin and with iron core is called ferritin. The iron core formed by the apoferritin contains about 4500 iron atoms.

We produced recombinant H- and L-apoferritins (rHF and rLF), which are homopolymers composed of H- and L-chain subunits from horse spleen apoferritin. Figure 6.3 shows the thermal stabilities of native ferritin, rHF, and rLF. Fer-8 is a rLF derivative which lacks the first eight residues from N-terminus, and Fer0 is the rLF. In the case of native ferritin, the heat capacity increased around 60 °C which indicates dissociation of the subunits starts as low as 60 °C. Compared to the native ferritin, recombinant ferritins' denaturation starts at higher temperatures. The lower subunit-dissociation temperature of the native ferritin may come from the situation that their subunits are heterogeneous and slight structural differences make the native ferritin less stable than homopolymer apoferritins. It was clearly indicated that the recombinant rLF has greater stability than rHF. The mutational studies indicated that both intra- and interchain contacts are important for ferritin stability (Yoshizawa et al. 2007). Based on thermal stabilities, we employed L-apoferritin (rLF) as a biotemplate to synthesize artificial cores, NPs.

We also used *Listeria innocua* Dps (DNA-binding protein from starved cells) as a biotemplate to synthesize smaller NPs. The protein shell is composed of 12 subunits, and the structure of the subunit was a 4-alpha-helix bundle and quite similar with ferritin subunit except the absence of helix E. Its outer and inner diameters are 9.5 nm and 4.5 nm, respectively (Bozzi et al. 1997; Su et al. 2005; Kornelius 2012). Like apoferritin, Dps has small channels connecting its cavity and outsides through which irons can be introduced. Dps can accumulate about 500 iron atoms within its cavity through a process facilitated by a ferroxidase center (Yoshizawa et al. 2007). By surveying appropriate solution conditions, NP can be synthesized artificially in the Dps cavity (Iwahori et al. 2007; Okuda et al. 2010).

6.3 Nanoparticle Synthesis Inside Ferritin and Dps Cavities

In preceding ferritin works, native apoferritins purified from tissues were commonly used (Harrison and Arosio 1996; Meldrum et al. 1991, 1992, 1995; Mann 1993; Douglas et al. 1995; Douglas and Stark 2000; Allen et al. 2003). We employed more stable recombinant apoferritins as a biotemplate, as well as native ferritin, horse spleen ferritin (HsAFr). We adopted a one-pot synthesis method, because the method is suitable for mass production due to its simplicity.

We first studied metal oxide/hydroxide NP biomineralization (Tsukamoto et al. 2005; Okuda et al. 2003). Apoferritin inner surface has areas of collective negatively charged amino acids which are putative nucleation sites. Since the electrostatic inner potential is lower than outsides (Takahashi et. al. 2003), positively charged ions, metal ions are easy to be introduced inside through the threefold channel, and ion concentration in the cavity is much higher than outside. As a result, metal oxide/hydroxide nuclei would be formed at the putative nucleation sites much faster than outside. Once the core is formed, the surface could work as self-catalytic way, and the metal ions would form selectively inside the cavity. Based on such assumption, we tried biomineralization of several kinds of metals and succeeded to synthesize metal complex cores, such as Fe, Cr, Co, and In oxide and Ni hydroxide.

One example of metal oxide NP synthesis is cobalt oxide (Co_3O_4). Cobalt oxide (Co_3O_4) NP was synthesized in HsAFr, rLF, and Fer8 (Tsukamoto et al. 2005). A throughout survey of optimum conditions for the Co_3O_4 core synthesis was carried out. 3 mM Co(II) ion, 0.5 mg/mL apoferritin, and 100 mM HEPES pH 8.3 buffer solution were added to hydrogen peroxide (H_2O_2) and kept at 50 °C. Upon H_2O_2 introduction, Co (II) ion started oxidization. After overnight incubation, Co_3O_4 cores were formed in over 90% of apoferritin cavities. The synthesized core analysis by X-ray photoemission spectroscopy (XPS) and electron energy-loss spectroscopy (EELS) proved that they contained cobalt atoms, and the X-ray powder diffraction (XRD) structure analysis confirmed the core structure as Co_3O_4. The elevated temperature was important to synthesize Co_3O_4.

Metal hydroxide NP is also successfully synthesized. Nickel hydroxide NPs were fabricated by mixing ammonium nickel sulfate solution with 0.3 mg/mL HsAFr and 5 mM ammonium nickel sulfate containing dissolved carbon dioxide (Okuda et al. 2003). The pH was kept at 8.65 by a mixture of 150 mM HEPES (pH 7.5) and 195 mM CAPSO (pH 9.5) with 20mM ammonium. Ammonium ions made nickel ions stable and suppress bulk precipitation. It was shown that the carbonate ions are indispensable for Ni hydroxide core formation. The core formation ratio was nearly 100%. The synthesized NPs were characterized by energy dispersive X-ray spectrometry (EDS), and nickel-specific peaks (7.5 KeV and 0.9 KeV) were observed. The synthesized core was Ni hydroxide.

After successful synthesis of metal complexes, we tried compound semiconductor NP synthesis. Wong et al. applied incremental addition of source ions and synthesized CdS NP in the cavity with diameters of 2.5 nm and 4.5 nm (Wong and Mann 1996). We newly designed a slow chemical reaction system (SCRY) for one-pot synthesis method and succeeded to synthesize compound semiconductors (Yamashita et al. 2004).

In the case of CdSe NP, synthesis by the SCRY was carried out as follows. Chemical reaction between Cd^{2+} and Se^{2-} is very fast in aqueous solution and produces CdSe aggregation very quickly. Cd^{2+} was first stabilized by excess ammonia. Four ammonias coordinate Cd^{2+} ion and form stable tetraaminecadmium ions ($Cd(NH_3)_4{}^{2+}$). Se^{2-} ions are supplied by selenourea which degrades very slowly and releases Se^{2-} for several hours. Even when Se^{2-} ions are supplied into the solution, tetraaminecadmium ions are so stable that CdSe synthesis is very slow in the bulk solution. On the other hand, when tetraaminecadmium releases positively charge Cd^{2+} ion near the threefold channel, Cd^{2+} ion had a chance to enter the cavity and was condensed at nucleation sites. When Se^{2-} ions enter the cavity by chance under such conditions, CdSe nuclei are instantaneously formed. Once the nuclei are made, the CdSe surface works as a self-catalyst, and homogeneous CdSe semiconductor NPs are synthesized in the cavity. The consumption of Cd^{2+} and Se^{2-} by the crystal growth in the cavity further lowers the nucleation rates in the bulk solution. As a result, chemical reaction speed difference between outside and cavity was so large that CdSe NP is only produced in the cavity.

An optimized reaction solution was 0.3 mg/ml HsAFr, 1 mM cadmium acetate, 1 mM selenourea, 7.5 mM ammonia water, and 40 mM ammonia acetate which produced homogeneous CdSe NP in the HsAFr cavity (Yamashita et al. 2004). XPS data confirmed that the NPs are composed of Cd and Se. After removal of the protein shell by 500° C heat treatment, a set of fairly sharp XRD peaks was observed, and the core structure was determined to be a cubic phase (zinc blend) CdSe with a little hexagonal phase (wurtzite). The average diameter of the obtained CdSe cores using TEM images was 7 nm, which is the same as that of the apoferritin cavity. The dispersion of diameters was small, and the standard deviation of the diameter was less than 10% of the average diameter. The SCRY can be applied in various

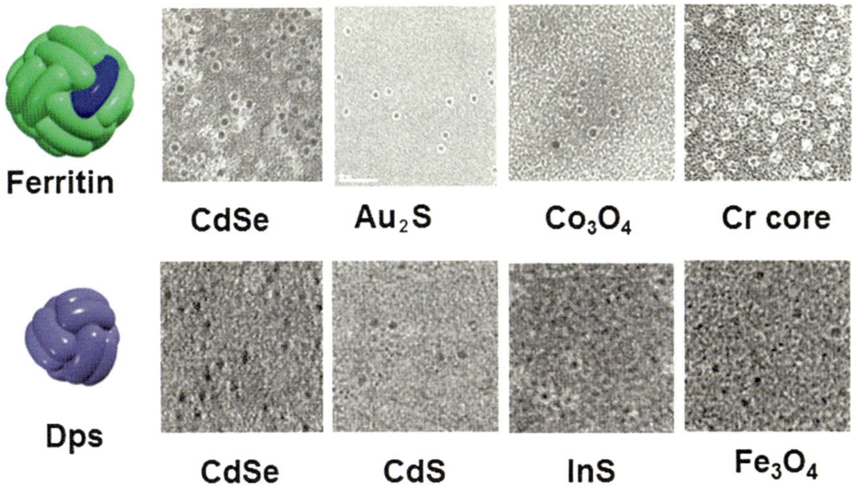

Fig. 6.4 TEM images of artificially synthesized NPs inside cage-shaped proteins. Images were negatively stained by 1% aurothioglucose which is too big to go through the channels and cannot stain the cavity

semiconductor NP syntheses. Homogenous ZnSe, CdS, ZnS, and NPs are successfully synthesized in the HsAFr and rLF. (Iwahori et al. 2005, 2009; Iwahori and Yamashita 2007, 2008)

We succeeded to mineralize lots of other materials inside cage-shaped protein. Among them were CdSe, Au_2S, Co_3O_4, Ta_2O_5, ZnO, $Cu(OH)_X$, $Ni(OH)_X$, InO_X, ZnSe, CdS, ZnS, CuS, AgS, and PtS inside ferritin and Fe_3O_4, Fe-Oxide, ZnO, $Cu(OH)_X$,$Ni(OH)_X$, InO_X, ZnS, CuS, and PtS inside Dps (Fig. 6.4).

6.4 Nanoparticle Arraying and Placement

For making nanoelectronics key-component as shown in Fig. 6.1, the synthesized NPs should be arrayed or placed on the designated area or positions. The protein shell of the cage-shaped protein plays an important role in this process. We employ two kinds of interactions. One is specific binding by peptide aptamer, and the other is controlled electrostatic interaction.

Aptamer is a target-specific binding peptide. We obtained aptamers by biopanning utilizing M13 phage display. M13 phage display combinatorial library of random 12 amino acid peptides, more than billions of different peptides at its end, were used. The phages were reacted with target inorganic materials. After washing by buffer solution, the M13 phages strongly bound to target material are eluted from the target surface and amplified by *E. coli*. This procedure was repeated several times to select the phage with the target-specific binding peptides. The genetic information for the 12 amino acids was analyzed, and its sequences were determined.

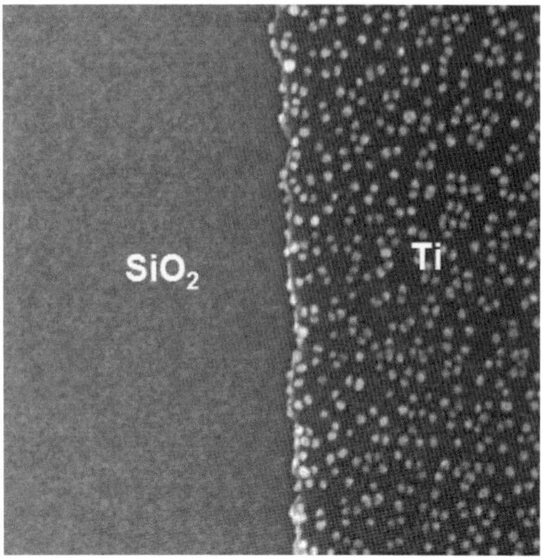

Fig. 6.5 SEM image of Au NPs selectively delivered to Ti oxide surface by the porter protein TFG. Ti oxide layer was formed left half by sputtering

In some cases, alanine scan was carried out, and unnecessary amino acids were eliminated from the sequences. We mainly used titanium (Ti)-binding peptide, carbonaceous material-binding peptide, and gold-binding peptides (Sano et al. 2005a; Kase et al. 2004; Brown 1997).

Fer-8 subunit was modified by a titanium (Ti)-binding peptide (TBP, MRKLPDA) at N-termini and gold-binding peptides (GBP) at the C-termini, named TFG. The subunits can make stable dimmers, and these dimers were mixed with 15–20 nm Au NP and incubated. Negatively stained TEM observation showed that Au NPs were dispersive and surrounded by the dimers. After purification, solution of Au NP-protein conjugates (TFG-Au) in 50 mM Tris-HCl pH 8 and 150 mM NaCl with 0.1–0.2% Tween 20 was put onto the substrate with Ti oxide pattern. The substrate was washed with pure water and dried. Figure 6.5 shows the SEM image of the TFG-Au conjugates on the substrate. The Au NPs were only distributed on the Ti oxide layer. TFG-Au NPs were closely packed, but NPs were dispersive because of the surrounding porter proteins. Plasmonic properties of Au NP were the same with original Au NPs, and AU NPs should be plasmonically insulated by the outer shell. This result clearly shows it is possible that protein with aptamers can catch specific nanomaterial and deliver them at the designated position or areas without aggregation (Zheng et al. 2010, 2011).

Another aptamer method was realized by mutant ferritins which displayed Ti-binding peptides on its outer surface. The mutant was proved to be able to biomineralize several kinds of nanomaterials as a core. The mutant ferritin showed strong affinity to the Ti oxide surface (Sano et al. 2005b; Yamashita et al. 2006), and naturally, core NP inside the cavity was delivered onto the Ti oxide patterns. A hexagonal 1.5-nm-thick Ti pattern with a diagonal length of 130 nm was made on a silicon substrate. A drop of solution of the mutant ferritin with NP was put on the substrate and, after 1 h, intensely washed by buffer and pure water. SEM observation

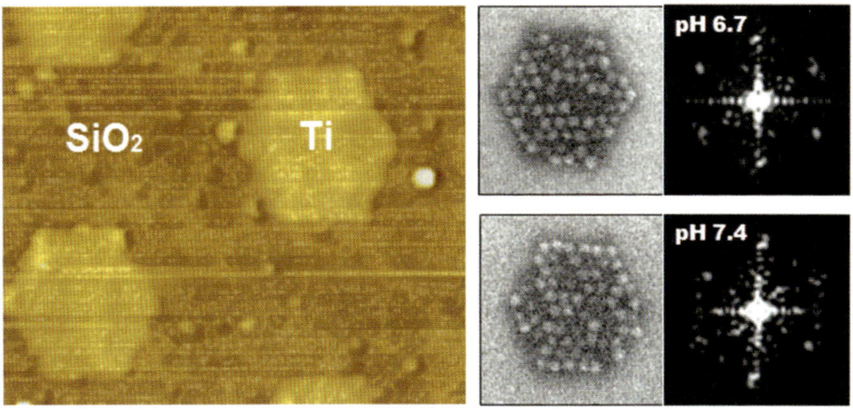

Fig. 6.6 AFM image of Ti hexagonal pattern (left). SEM image of mutant ferritin adsorption on the Ti pattern at pH 6.7 and pH 7.4 and fast Fourier transform of the corresponding SEM image (right). White dots represent the accommodated NPs

clearly showed that the mutant ferritins were bound only on the Ti pattern. Moreover fast Fourier analysis indicated that ferritins are hexagonally close-packed (Fig. 6.6) (Matsukawa et al. 2009). This simple delivery and array method is a promising method to make new functions such as mini-band NP structures.

We also modified the ferritin subunit to have carbonaceous material-binding peptide which binds to carbon nano-horn materials (Kase et al. 2004). Aptamer against carbonaceous materials (NHBP-1) was genetically attached at the N-termini of the subunit, and the mutant ferritin could bind to carbon nanotubes. Judging from its sequence, they are hydrophobic; even the mutant ferritins are well-soluble in the aqueous solution. The well-dispersive characteristic could be due to the electrostatic repulsive force among mutant ferritins which are net negatively charged around neutral pH. It was expected that the mutant ferritins would have attractive forces among them when they are enforced to be close enough for aptamers to interact each another. Based on this assumption, we prepared Si substrate covered by polyethylene glycol (PEG, MW 2000 or 4000) with the spin-coat technique. PEG layer is soft and allows mutant ferritins to move slightly in lateral direction after adsorption, which will lead to rearrangement of the ferritins and formation of 2D crystalline array (Tsukamoto et al. 2012).

Mutant ferritin solution was put on the PEG-coated substrate, and excessive solution was spun out. Figure 6.7 shows a SEM image of ferritin adsorption on PEG-coated SiO_2. White dots represent the accommodated NPs. Several hexagonal closely packed ferritin array domains with diameters of larger than 200 nm were assembled. The FFT result from $200 \times 200\ nm^2$ area showed six peaks corresponding to a hexagonally close-packed NP array. The mutant ferritins were pushed down to the PEG-coated substrate during the spin-dry and condensed. During being two-dimensionally confined on the substrate, short-range hydrophobic attractive force might have made them form 2D crystalline array (Tsukamoto et al. 2012).

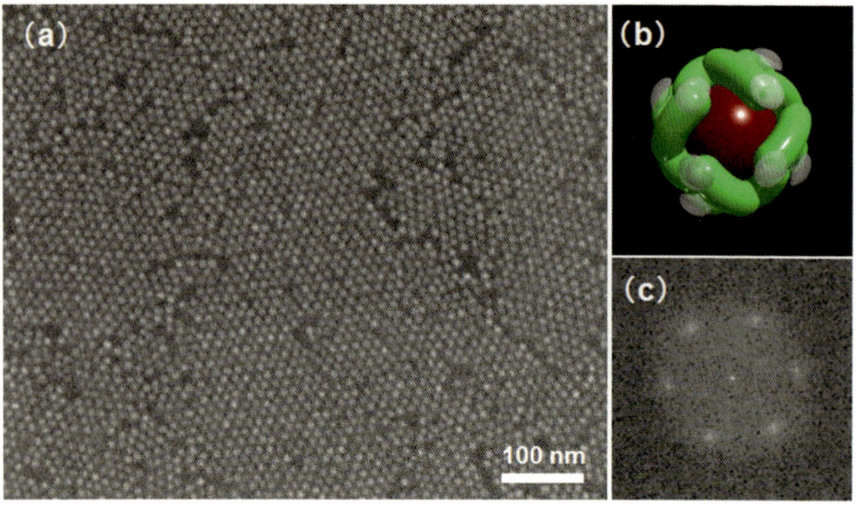

Fig. 6.7 SEM image of ferritin adsorption on PEG-coated SiO_2 substrate (**a**), the schematic drawing of apoferritin with flappy carbonaceous material-binding peptides without four foreground subunits (**b**), and the fast Fourier transfer (FFT) result from 200 × 200 nm^2 area (**c**)

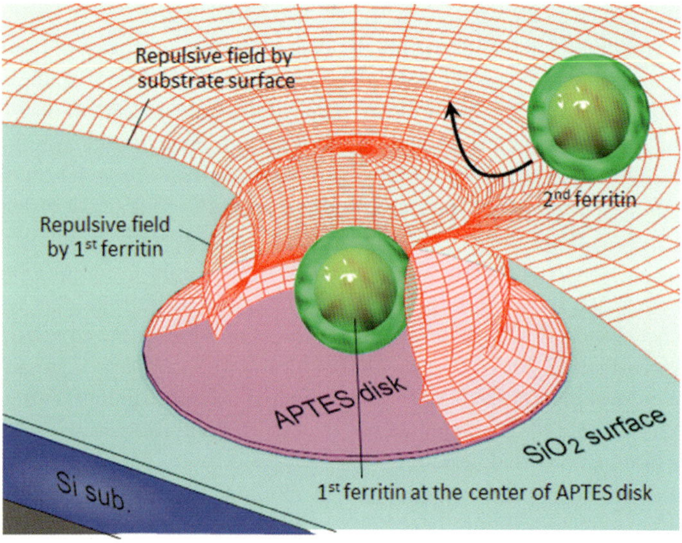

Fig. 6.8 Schematic for single ferritin adsorption onto APTES disk. The negative charge-enhanced ferritin adsorbs onto the APTES disk center, and the second approaching ferritin will be repelled

The controlled electrostatic interaction is another method to handle the NPs. Precise control of electrostatic interaction between net negatively charged cage-shaped protein and nanometric positively charged pattern on the Si substrate can deliver cage-shaped protein with NP in a one-by-one manner (Fig. 6.8) (Yoshii et al.

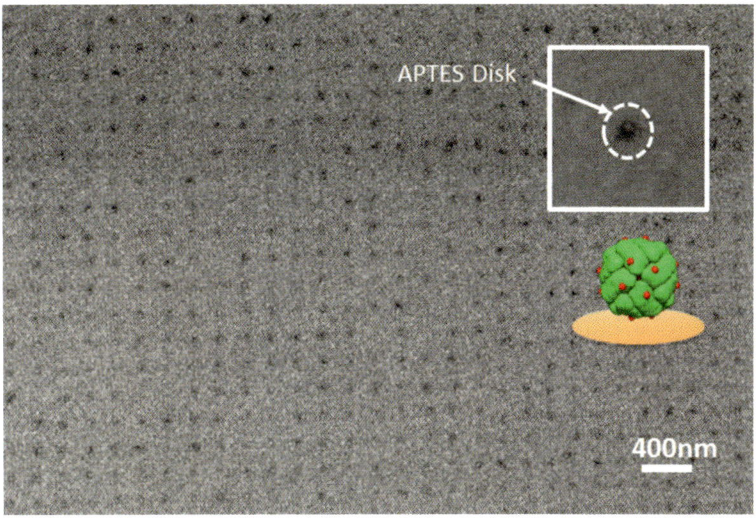

Fig. 6.9 SEM image of a single mutant ferritin on a 45 nm APTES disk. Iron oxide NPs in ferritins are observed as bright white dots, where APTES disks are observed as dark areas. Each APTES disk has single ferritin molecule. Red marks on a ferritin show amino acid replacement points (inset)

2009; Sano et al. 2005c). Ferritin outer surface has negatively and positively charged amino acids, and its net charge is negative around neutral pH. We enhanced the surface more negative by genetically replacing lysine residues to glutamic acids. We produced positive circular 3-aminopropyl-triethoxysilane (APTES) patterns with 45 nm of diameter on Si substrate surface which was oxidized and highly negative. Numerical Poisson-Boltzmann equation calculations were carried out under a variety of conditions to generate a three-dimensional electrostatic potential showing a funnel-shaped path for ferritin to reach the positive nanodisk center. Under such conditions, it was expected that the first adsorbed ferritin cancels APTES-positive potential and another ferritin cannot approach the APTES disk.

The simulation indicated that under appropriate ionic strength, i.e., Debye length, single mutant ferritin with NP can be adsorbed on an APTES disk. Following the simulation, adsorption experiment was carried out, and Fig. 6.9 is the result. The SEM image clearly showed that the ferritin with NP is positioned at the APTES disk with 200 nm intervals, one by one. It was confirmed that precisely controlled electrostatic interaction could place a single ferritin on an APTES disk. The delivered independent NPs were used as catalysts for carbon nanotube growth and metal-induced lateral crystallization (MILC). There are also many other applications of the independent NP array.

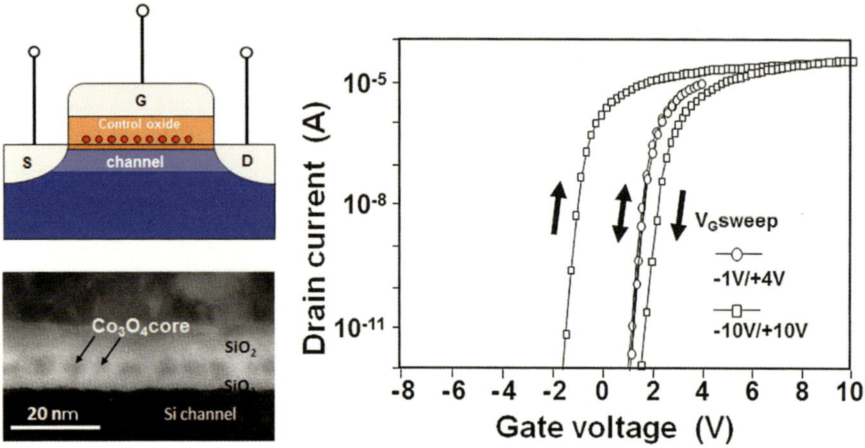

Fig. 6.10 Schematic drawing of floating nanodot gate memory (FNGM) and the cross-sectional TEM image (left). I_D-V_G characteristics of FNGM with Co-NPs synthesized and arrayed by ferritin (right)

6.5 Some Applications

6.5.1 Floating Nanodot Gate Memory Applications of NPs Produced and Arrayed by Ferritin Molecules

The introduction of the BNP into the Si process was a big challenge. It was out of question to introduce biomaterials into the semiconductor process and bring them into clean rooms. We first applied the BNP to produce the charge storage nodes of floating nanodot gate memory (Miura et al. 2006). Floating nanodot gate memory (FNGM) is a MOS-FET equipped with NP array or charge storage nodes in the control oxide layer above the MOS channel. Ferrihydrite and Co_3O_4 NPs were synthesized in apoferritin or Dps. The obtained samples were put on a size exclusion chromatography with salt-free buffer, which produced the salt-free ferritin/Dps solution. This solution contained less than 50 ppm sodium and could be introduced in Si process. Using interaction of the outer protein shells and substrate, they are two-dimensionally arrayed on a thin tunnel SiO_2 layer above the channel. After protein elimination by heat treatment at 500 °C for 10 min in oxygen gas or UV/ozone treatment, which was confirmed by AFM, FTIR, and XPS measurements, the NPs were buried in the control thick oxide layer. In the case of a two-dimensional array, the density was around 5–8 × 10^{11}/cm^2 for ferritin and up to 1.8 × 10^{12}/cm^2 for Dps. After electrodes had been deposited, post-metal annealing under forming gas (H_2/N_2, 10%/90%) was carried out at 450 °C for 1 h. TEM observation clearly showed that NP retained spherical shape and dispersed. No aggregation was observed. Each NP was embedded completely surrounded by SiO_2 (Fig. 6.10).

The I_D-V_G curves of the fabricated FNGM showed that a narrow gate voltage sweep produced the typical I_D-V_G curve of a normal n-channel MOS-FET. When the gate voltage swept from −10 V and +10 V, the I_D-V_G curves showed a hysteresis and threshold voltage (V_{th}) at which I_D shuts off, depending on the sweep direction (Miura et al. 2006). When positive gate voltage was high enough, the electrons in the n-channel tunneled into the NP array and shut off the drain current at the higher gate voltage. On the other hand, when the negative gate voltage was low enough, holes were stored in the NP array and worked in reverse. This hysteresis demonstrated that the FNMG with NPs produced and arrayed by apoferritin can work as a good memory. The endurance test of the FNGM confirmed that degradation was negligible up to 10^5 cycles of program/erase operation. The retention time or charge confinement time was measured up to 10,000 s. It was confirmed that the charge can be stored for practical lengths of time (Miura et al. 2006).

This is the first legitimate introduction of proteins into the Si process. The BNP can also produce one-dimensional array of NPs by modifying the adsorption conditions. This one-dimensional NP array was applied to ultrashort junctionless transistor. The produced transistor showed ultra-large hysteresis and good memory effect (Ban et al. 2015). The BNP for FGNM is further developing using ferritin and Dps and has been improving (Yamada et al. 2007; Ban et al. 2015; Miura et al. 2008; Kamitake et al. 2015).

6.5.2 Carbon Nanotube (CNT) Synthesis Using Cage-Shaped Protein

Since NPs synthesized in ferritin and Dps are nanometric and homogeneous, the NPs are suitable catalysts after protein elimination (Kim et al. 2001; Kramer et al. 2005; Jeong et al. 2005; Takagi et al. 2007; Kumagai et al. 2010; Sugano et al. 2016). Dai et al. synthesized CNTs using ferritin (Kim et al. 2001). Naik et al. used Dps to produce CNTs (Kramer et al. 2005). In addition to the homogeneity, Kumagai et al. used the position-controlled ferritin placement technique to grow CNTs. Periodically placed single ferritin, utilizing the same process shown above, made NP array with 50 nm interval after protein elimination. The NPs worked as catalysts, and CNT grew vertically to have made a CNT forest (Kumagai et al. 2010). Isono et al. used CNT forest grown from ferritin iron oxide core array to fabricate micromechanical resonator-type gas analyzer. They selectively placed ferritins with iron oxide cores on the resonator, and after protein elimination, they produced CNT forest using the iron oxide cores as catalysts. The density and location of CNT forest were well-controlled. The produced analyzer with CNT forest as gas adsorber showed clear sensitivity (Fig. 6.11) (Sugano et al. 2016).

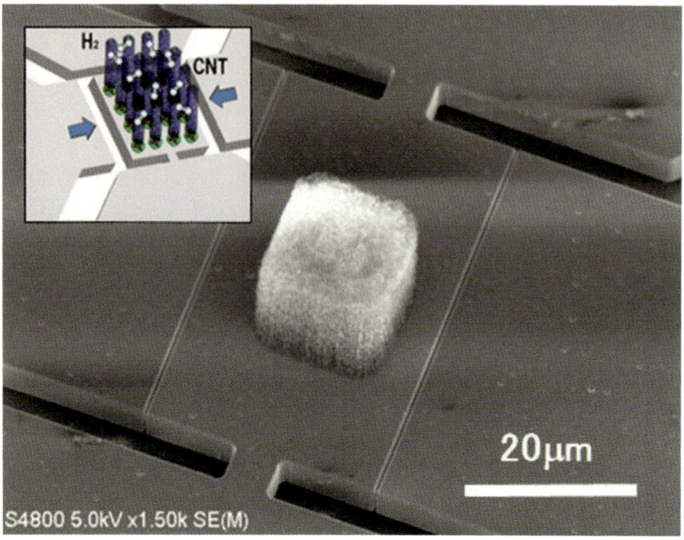

Fig. 6.11 SEM image of micromechanical resonator-type gas analyzer. CNTs were selectively grown on the central resonator which worked as gas absorber. *Inset* is the schematic drawing of the resonator and CNTs

6.5.2.1 Bioconjugate of CNT and Dps

The BNP can construct not only two-dimensional (2D) functional structures on semiconductor substrate but also three-dimensional (3D) structures. One of the 3D structures by BNP is a bioconjugate of carbon nanotubes (CNTs) and Dps protein with inner NP. We genetically modified the Dps by attaching carbonaceous material-binding peptides (NHBP-1) to the N-terminus of the Dps subunit. Mutant proteins expressed in *E. coli* were found to self-assembled into cage-shaped protein, named NHBP-Dps. NHBP-Dps could form NPs of iron oxide, compound semiconductors, Co_3O_4, and so on in its cavity. Mixing the NHBP-Dps with NP and CNTs produced a nanostructure where CNTs were surrounded by NHBP-Dps and the NHBP-Dps caught other CNTs (Kobayashi et al. 2011). A NHBP-Dps with NP was sandwiched by two CNTs. The NP inside could be a quantum dot, and protein shell could form tunnel junctions. Soft attachment by the aptamers realized that intact CNT properties were preserved.

Electronic properties of the single-wall carbon nanotubes (SWNTs) conjugated with NHBP-Dps with Co_3O_4 NP were characterized (Kumagai et al. 2012). The NHBP-Dps (Co_3O_4) was mixed with the SWNTs in 20 mM HEPES-NaOH (pH 7.5) in a final concentration of 0.3 mg/mL each. Ultrasonication was applied for 5 min, and the solution was centrifuged at 15,000 rpm for 10 min at 4 °C. TEM observation of the supernatant showed SWNTs were covered by the NHBP-Dps (Co_3O_4). The same nanostructures could be produced using ferritin molecules. Solution of SWNTs conjugated with NHBP-Dps (Co_3O_4) was dropped onto a pair of Au/Cr electrodes with 1–2.5 μm gap. 1 MHz 6V for 5–10 s was applied to align the SWNT

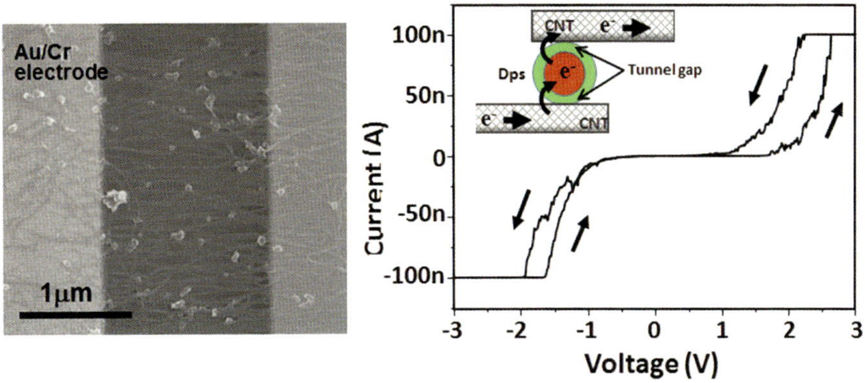

Fig. 6.12 SEM imaged of the conjugate material at electrode gap (left). Current-voltage characteristics of the obtained conjugate (right). Inset is a schematic drawing of tunnel junction formed in SWNTs conjugated with NHBP-Dps (Co_3O_4)

parallel to the electric field with dielectrophoresis effect. The sample was baked at 110 °C for 5 min and dried. Figure 6.12 shows the current-voltage characteristics of the buffer-free bioconjugate network. The coulomb blockade suppressed current at low bias voltage. It was suggested that the formation of conductive paths as a result of Co oxide reduction within NPs caused the hysteresis. The results indicated that the bioconjugates can be a new type of 3D hetero-material and novel properties generated by the conjugation were expected.

6.5.3 Bioconjugate of CNT and Dps for Thermoelectric Device Application

Nakamura et al. applied the bioconjugate of CNT and Dps to thermoelectric device (Fig. 6.13) (Ito et al. 2014). CNTs have mechanically high strength, lightness, and high electric conductivity and are expected to produce flexible thermoelectric devices. The CNTs, however, have a large thermal conductivity and fairly low Seebeck coefficient. CNT-Dps bioconjugates could overcome these weak points. CNTs were mixed with NHBP-Dps with semiconductor core in solution under appropriate conditions, and CNTs were covered partially by NHBP-Dps. The obtained conjugates were condensed and observed by TEM. The condensed black material had many CNT/NHBP-Dps/CNT junctions. It was predicted that phonon is scattered at the CNT/soft-shell interface which suppresses heat flow of the conjugates and reduces the thermal conductivity significantly. This leads to generate a stepwise temperature difference across the junction. The large temperature difference generates an asymmetric Fermi-Dirac distribution of electrons/holes in the both side CNTs. This asymmetry cause unidirectional electrons/holes flow through the conduction/valence band of the core. This phenomenon is a new type of Seebeck effect.

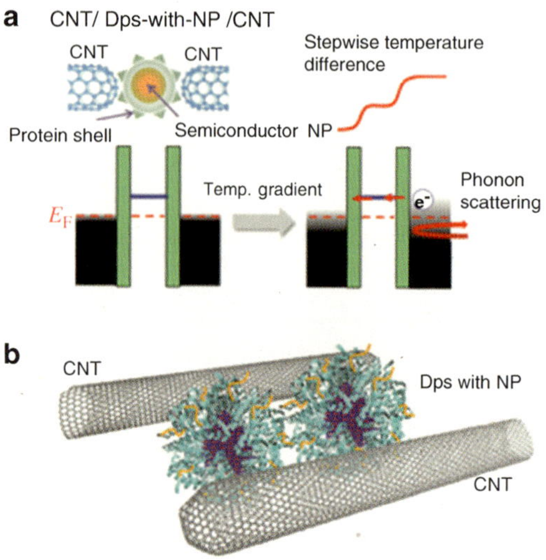

Fig. 6.13 Schematic drawings of (**a**) Seebeck effect at a biomolecular junction, (**b**) drawing of two C-Dps molecules bridging two CNTs. Note that CNTs are enlarged for ease (Copyright (2014) The Japan Society of Applied Physics)

Table 6.1 Thermoelectric properties of pristine CNT, CNT/Dps(Fe), and CNT/Dps(CdSe)

	Thermal conductivity (m/Wk)	Seebeck coefficient (μV/K)	Electrical conductivity (S/cm)	Relative ZT (at 300 K)
CNT(pristine)	17.2 ± 0.47	33.1	0.30	1
CNT/Dps(Fe)	2.0 ± 0.03	56.3 ± 11.5	0.84 ± 0.07	1071
CNT/Dps(CdSe)	–	28.0 ± 5.4	0.21 ± 0.11	(66)

The conjugate films of NHBP-Dps with iron oxide NP and CNTs were produced, and their thermal and electrical properties were measured. Thermal conductivity was markedly suppressed and Seebeck coefficient increased. As a result, the figure of merit, $ZT = \alpha^2\sigma T/\kappa$ (α,σ,κ and T are the Seebeck coefficient, electrical conductivity, thermal conductivity, and absolute temperature, respectively), is increased over 1000-fold. They also made the conjugates with NHBP-Dps with CdSe core, and the ZT was calculated to be increased around 66-fold (see Table 6.1) (Ito et al. 2014). The results demonstrated that the BNP can produce nanostructures for new-type thermoelectric materials easily under ecofriendly conditions.

6.6 Conclusion

It was demonstrated that the proposed bio nano process (BNP) can produce several kinds of homogenous NPs inside cage-shaped protein. Under the biotemplated biomineralization, the same size NP can be produced, and the surrounding protein shell can deliver the accommodated NPs at designated positions and locations. Site-specific binding peptides anchor the NPs at designed areas/positions. Precise control of electrostatic interaction can place a single NP onto an APTES nanodisk formed on substrate. The BNP can produce and place a variety of NPs. These NPs realize a variety of functions such as charge storage nodes for floating nanodot gate memory, catalysts for carbon nanotube growth, a quantum well in heterogeneous junctions, and nanoetching masks. It is proved that the obtained NP array is able to fabricate key components of nanoelectronic devices through the integration of top-down and bottom-up technology. As shown in this paper, the BNP is capable of producing functional nanostructures otherwise impossible by other methods.

The BNP is now proving the usefulness of biomolecules' bottom-up nanotechnology in the nanoelectronic device fabrication and that the integration of top-down and bottom-up technology is a powerful tool. The BNP is expanding its applications. This biological approach producing and handling NPs is now opening up a new path toward the fabrication of functional nanostructures realizing new devices which operate on completely new mechanisms.

References

Allen M, Willits D, Young M, Douglas T (2003) Constrained synthesis of cobalt oxide nanomaterials in the 12-subunit protein cage from Listeria innocua. Inorg Chem 42:6300–6305

Ban T, Uenuma M, Migita S, Okamoto N, Ishikawa Y, Yamashita I, Uraoka Y (2015) Ultra-short channel junctionless transistor with a one-dimensional nanodot array floating gate. Appl Phys Lett 106:253104

Bozzi M, Mignogna G, Stefanini S, Barra D, Longhi C, Valenti P, Chiancone E (1997) A novel non-heme iron- binding ferritin related to the dna- binding proteins of the dps family in Listeria innocua. J Biol Chem 272:3259–3326

Brown S (1997) Metal-recognition by repeating poly peptides. Nat Biotechnol 15:269–272

Clegg GA, Fitton JE, Harrison PM, Treffry A (1980) Ferritin: molecular structure and iron storage mechanisms. Prog Biophys Mol Biol 36:56–86

Douglas T, Stark VT (2000) Nanophase cobalt oxihydroxide mineral synthesized within the protein cage of ferritin. Inorg Chem 39:1828–1830

Douglas T, Dickson DPE, Betteridge S, Charnock J, Garner CD, Mann S (1995) Synthesis and structure of an iron(III) sulfide-ferritin bioinorganic nanocomposite. Science 269:54–57

Gallois B, d'Estaintot BL, Michaux MA, Dautant A, Granier T, Précigoux G, Soruco JA, Roland F, Chavas-Alba O, Herbas A, Crichton RR (1997) X-ray structure of recombinant horse l-chain apoferritin at 2.0 Å resolution: implications for stability and function. J Biol Inorg Chem 2(360–367)

Harrison M, Arosio P (1996) The ferritins: molecular properties, iron storage function and cellular regulation. Biochem Biophys Acta 1275:161–203

Harrison PM, Andrews SC, Artymuik PJ, Ford GC, Guest JR, Hirzmann J, Lawson DM, Livingstone JC, Smith JMA, Treffry A, Yewdall SJ (1991) Probing structure-function relations in ferritin and bacterioferritin. Adv Inorg Chem 36:449–486

Ito M, Okamoto N, Abe R, Kojima H, Matsubara R, Yamashita I, Nakamura M (2014) Enhancement of thermoelectric properties of carbon nanotube composites by inserting biomolecules at nanotube junctions. Appl Phys Express 7:065102

Iwahori K, Yamashita I (2007) Fabrication of CdS nanoparticles in the bio-template, apoferritin cavity by a slow chemical reaction system. J Phys 61:492–496

Iwahori K, Yamashita I (2008) Size-controlled one-pot synthesis of fluorescent cadmium sulfide semiconductor nanoparticles in an apoferritin cavity. Nanotechnology 19:495601

Iwahori K, Yoshizawa K, Muraoka M, Yamashita I (2005) Fabrication of ZnSe nanoparticles in the apoferritin cavity by designing a slow chemical reaction system. Inorg Chem 44:6393–6400
Iwahori K, Enomoto T, Furusho H, Miura A, Nishio K, Mishima Y, Yamashita I (2007) Cadmium sulfide nanoparticle synthesis in Dps protein from Listeria innocua. Chem Mater 19(13):3105–3111
Iwahori K, Furusho H, Yamashita I (2009) Characterization of ZnS compound semiconductor nano-particle in an protein cage, apoferritin, ICMAT 2009 & IUMRS-ICA 2009, Singapore, June 28–July 3, 2009
Jeong G-H, Yamazaki A, Suzuki S, Yoshimura H, Kobayashi Y, Homma Y (2005) Cobalt-filled Apoferritin for suspended single-walled carbon nanotube growth with narrow diameter distribution. J Am Chem Soc 127(23):823
Kamitake H, Uenuma M, Okamoto N, Horita M, Ishikawa Y, Yamashita I, Uraoka Y (2015) Floating gate memory with charge storage dots array formed by Dps protein modified with site-specific binding peptides. Nanotechnology 26:195201
Kase D, Kulp JL, Yudasaka M, Evans JS, Iijima S, Shiba K (2004) Affinity selection of peptide phage libraries against single-wall carbon nanohorns identifies a peptide aptamer with conformational variability. Langmuir 20(20):8939–8941
Kim YLW, Zhang Y, Rolandi M, Wang D, Dai H (2001) Growth of single-walled carbon nanotubes from discrete catalytic nanoparticles of various sizes. J Phys Chem B 105(46):11424–11431
Kobayashi M, Kumagai S, Zheng B, Uraoka Y, Douglas T, Yamashita I (2011) A water-soluble carbon nanotube network conjugated by nanoparticles with defined nanometre gaps. Chem Commun 47:3475–3477
Kornelius Z (2012) Dps biomineralizing proteins: multifunctional architects of nature. Biochem J 445:297–311
Kramer RM, Sowards LA, Pender MJ, Stone MO, Naik RR (2005) Constrained iron catalysts for single-walled carbon nanotube growth. Langmuir 21(18):8466–8470
Kumagai S, Ono T, Yoshii S, Kadotani A, Tsukamoto R, Nishio K, Okuda M, Yamashita I (2010) Position-controlled vertical growths of individual carbon nanotubes using a cage-shaped protein. Appl Phys Express 3:015101
Kumagai S, Okamoto N, Kobayashi M, Yamashita I (2012) Characterisation of a carbon nanotube random network conjugated by semiconductor nanoparticles with defined nanometre-scaled gaps. Micro Nano Lett 7(8):753–756
Lawson DM, Artymiuk PJ, Yewdall SJ, Smith JMA, Livingstone JC, Treffry A, Luzzago A, Levi S, Arosio P, Cesareni G, Thomas CD, Shaw WV, Harrison PM (1991) Solving the structure of human H ferritin by genetically engineering intermolecular crystal contacts. Nature 349(541–544)
Levi S, Luzzago A, Cesareni G, Cozzi A, Franceschinelli F, Albertini A, Arosio P (1988) Mechanism of ferritin iron uptake: activity of the H-chain and deletion mapping of the ferro-oxidase site. A study of iron uptake and ferro-oxidase activity of human liver, recombinant H-chain ferritins, and of two H-chain deletion mutants. J Biol Chem 263(18086–18092)
Mann S (1993) Molecular tectonics in biomineralization and biomimetic materials chemistry. Nature 365:499–505
Massover WH (1993) Ultrastructure of ferritin and apoferritin – a review. Micron 24:389–437
Matsukawa N, Nishio K, Sano K, Shiba K, Yamashita I (2009) Hexagonal close-packed array formed by selective adsorption onto hexagonal patterns. Langmuir 25(6):3327–3330
Meldrum FC, Wade VJ, Nimmo DL, Heywood BR, Mann S (1991) Synthesis of inorganic nanophase materials in supramolecular protein cages. Nature 349:684–687
Meldrum FC, Heywood BR, Mann S (1992) Magnetoferritin: in vitro synthesis of a novel magnetic protein. Science 257:522–523
Meldrum FC, Douglas T, Levi S, Arosio P, Mann S (1995) Reconstitution of manganese oxide cores in horse spleen and recombinant ferritins. J Inorg Biochem 58:59–68
Miura A, Hikono T, Matsumura T, Yano H, Hatayama H, Uraoka Y, Fuyuki T, Yoshii S, Yamashita I (2006) Floating nanodot gate memory devices based in biomineralized inorganic nanodot array as a storage node. Jpn J Appl Phys 45:L1–L3

Miura A, Tsukamoto R, Yoshii S, Yamashita I, Uraoka Y, Fuyuki T (2008) Non-volatile flash memory with discrete bionanodot floating gate assembled by protein template. Nanotechnology 19:255201

Okuda M, Iwahori K, Yamashita I, Yoshimura H (2003) Fabrication of nickel and chromium nanoparticles using the protein cage of apoferritin. Biotechnol Bioeng 84:187–193

Okuda M, Suzumoto Y, Iwahori K, Kang S, Uchida M, Douglas T, Yamashita I (2010) Bio-templated CdSe nanoparticle synthesis in a cage shaped protein, Listeria-Dps, and their two dimensional ordered array self-assembly. Chem Commun 46:8797–8799

Sano K, Sasaki H, Shiba K (2005a) Specificity and biomineralization activities of Ti-binding peptide-1 (TBP-1). Langmuir 21:3090–3095

Sano KK, Ajima K, Iwahori M, Yudasaka S, Iijima I, Yamashita S, Shiba K (2005b) Endowing a ferritin-like cage protein with high affinity and selectivity for certain inorganic materials. Small 1(8–9):826–832

Su M, Stefanini S, Chiancone E, Chasteen ND (2005) The so-called *Listeria innocua* ferritin is a Dps protein. Iron incorporation, detoxification, and DNA protection properties. Biochemistry 44(15):5572–5578

Sugano K, Matsumoto R, Tsutsui R, Kishihara H, Matsuzuka N, Yamashita I, Uraoka Isono YY (2016) Fabrication and characterization of a CNT forest integrated micromechanical resonator for a rarefied gas analyzer in a medium vacuum atmosphere. J Micromech Microeng 26:075010

Takagi D, Yamazaki A, Otsuka Y, Yoshimura H, Kobayashi Y, Homma Y (2007) Gold-filled apoferritin for investigation of single-walled carbon nanotube growth on substrate. Chem Phys Lett 445:213–216

Takahashi T, Kuyucak S (2003) Functional properties of threefold and fourfold channels in ferritin deduced from electrostatic calculations. Biophys J 84(4):2256–2263

Tsukamoto R, Iwahori K, Muraoka M, Yamashita MI (2005) Synthesis of Co_3O_4 nanoparticles using the cage-shaped protein, apoferritin. Bull Chem Soc Jpn 78(11):2075–2081

Tsukamoto R, Igarashi M, Samukawa S, Yamashita I (2012) Fast two-dimensional ferritin crystal formation realized by mutant ferritin and poly(ethylene glycol) modified SiO2 substrate. Appl Phys Express 5:065201

Wong KKW, Mann S (1996) Biomimetic synthesis of cadmium sulfide-ferritin nanocompsites. Adv Mater 8:928–932

Yamada K, Yoshii S, Kumagai S, Miura A, Uraoka Y, Fuyuki T, Yamashita I (2007) Effects of dot density and dot size on charge injection characteristics in nanodot array produced by protein supramolecules. Jpn J Appl Phys 46:7549–7553

Yamashita I (2000) Fabrication of a two-dimensional array of nano-particles using ferritin molecule. Thin Solid Films 393:12–18

Yamashita I, Hayashi J, Hara M (2004) Bio-template synthesis of uniform CdSe nanoparticles using cage-shaped protein, apoferritin. Chem Lett 33(9):1158–1159

Yamashita I, Kirimura H, Okuda M, Nishio K, Sano K, Shiba K, Hayashi T, Hara M, Mishima Y (2006) Selective nanoscale positioning of ferritin and nanoparticles by means of target-specific peptides. Small 2(10):1148–1152

Yoshii S, Kumagai S, Nishio K, Kadotani A, Yamashita I (2009) Electrostatic self-aligned placement of single nanodots by protein supramolecules. Appl Phys Lett 95:133702

Yoshizawa K, Mishima Y, Park S-Y, Heddle JG, Tame JRH, Iwahori K, Kobayashi M, Yamashita I (2007) Effect of N-terminal residues on the structural stability of recombinant horse L-chain apoferritin in an acidic environment. J Biochem 142:707–713

Zheng B, Yamashita I, Uenuma M, Iwahori K, Kobayashi1 M, Uraoka1 Y (2010) Site-directed delivery of ferritin-encapsulated gold nanoparticles. Nanotechnol 21:045305

Zheng B, Zettsu N, Fukuta M, Uenuma M, Hashimoto T, Gamo K, Uraoka Y, Yamashita I, Watanabe H (2011) Versatile protein-based bifunctional nano-systems (encapsulation and directed assembly) selective nanoscale positioning of gold nanoparticle-viral protein hybrids. Chem Phys Lett 506:76–80

Part III
Bio-Inspired Magnetic Materials for Nanotechnology-Based Applications

Chapter 7
Biogenic and Bio-inspired Syntheses of Hierarchically Structured Iron Compounds for Lithium-Ion Batteries

Hiroaki Imai

Abstract This chapter shows biogenic and bio-inspired syntheses of hierarchically structured iron compounds for lithium-ion batteries. Iron-reducing bacteria produce spherical microparticles consisting of iron(II) phosphate nanosheets. Lithium iron phosphate ($LiFePO_4$) with a hierarchical morphology including a high specific surface area is obtained from biogenic iron(II) phosphate via a hydrothermal reaction. The biogenic iron(II) phosphate and microbially derived $LiFePO_4$ show reversible charge-discharge performance as anodes and cathodes of lithium-ion batteries, respectively. The precursors of γ- and α-FeOOH with a high specific surface area are prepared through a microbial-mineralization-inspired approach in aqueous solutions. Subsequent thermal treatment under tuned conditions facilitates the syntheses of γ- and α-Fe_2O_3 without the collapse of the hierarchical structures and the high specific surface area. The resultant iron compounds are applicable for the electrodes of lithium-ion batteries, adsorbents, and catalysts.

7.1 Introduction

Biominerals produced by organisms possess hierarchical architectures composed of inorganic-organic composite materials (Gao et al. 2003; Young et al. 1999; Hamm et al. 2003). Their sophisticated architectures are valuable for practical usage as various functional materials. Several microorganisms produce transition metals or metal compounds by the oxidation or reduction of metal ions (Klaus et al. 1999; Konishi et al. 2006; Labrenz et al. 2000; Banfield et al. 2000; Hashimoto et al. 2007; Gorby et al. 1988; Arakaki et al. 2003; Villalobos et al. 2003). Specifically, biogenic iron compounds are produced by iron-oxidation bacteria including magnetotactic bacteria (Banfield et al. 2000; Hashimoto et al. 2007; Gorby et al. 1988; Arakaki et al. 2003). The biogenic iron oxides are known to provide good performance as

H. Imai (✉)
Department of Applied Chemistry, Faculty of Science and Technology, Keio University, Yokohama, Japan
e-mail: hiroaki@applc.keio.ac.jp

T. Matsunaga et al. (eds.), *Biological Magnetic Materials and Applications*,
https://doi.org/10.1007/978-981-10-8069-2_7

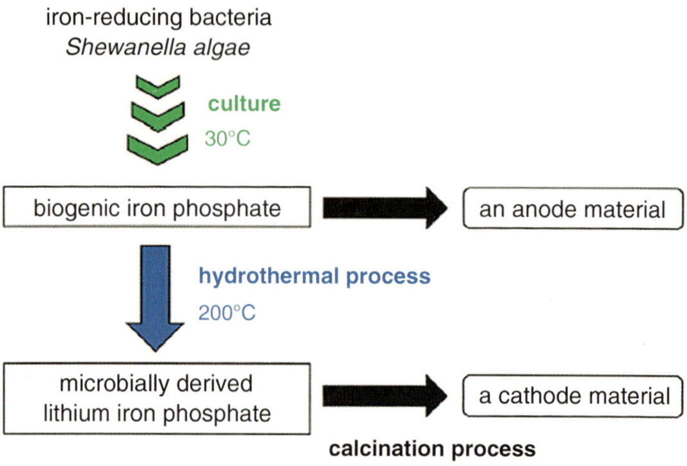

Scheme 7.1 Experimental procedure for the production of biogenic iron(II) phosphate and microbially derived lithium iron phosphate (Reproduced from Kageyama et al. 2015 with permission (Copyright 2015 The Royal Society of Chemistry))

functional materials. The application of the biogenic iron compounds has been studied for polymerization catalysis (Kanazawa et al. 2012) and enzyme carriers (Ema et al. 2011). Moreover, amorphous iron-based oxide nanoparticles produced by aquatic bacteria show potential as an anode material for lithium-ion batteries (Hashimoto et al. 2014).

Many metal compounds have been applied as anode and cathode materials for lithium-ion batteries (Reddy et al. 2013). Iron oxide (Thackeray et al. 1982; Larcher et al. 2003; Armand and Tarascon 2008; Cherian et al. 2012; Taberna et al. 2006; Reddy et al. 2007; Das et al. 2013; Petnikota et al. 2015), iron nickel oxide (Cherian et al. 2013), sulfide (Wu et al. 2011; Bernardi and Newman 1987) and phosphate (Son et al. 2004) were studied as anode materials owing to the electrochemical conversion reaction to lithium ions. As cathode materials of iron(II) compounds for lithium-ion batteries, lithium iron phosphate ($LiFePO_4$) has a high theoretical capacity and high thermal stability (Padhi et al. 1997; Yuan et al. 2011). Morphological control of the cathode materials is very important to enhance the interdiffusion of the lithium ion and electron transfer. A variety of hierarchical structures in anode and cathode materials including $LiFePO_4$ were synthesized by bio-templating methods (Nam et al. 2006; Xia et al. 2011, 2012, 2013; He et al. 2013; Chang et al. 2014; Shim et al. 2015; Tao et al. 2014a, b, 2015).

In the first part of this chapter, biogenic minerals produced by a microorganism are described as useful practical anode and cathode materials with a particular nanostructure (Kageyama et al. 2015). As shown in Scheme 7.1, biogenic iron(II) phosphate was produced by the iron-reducing bacteria *Shewanella algae*, and microbially derived $LiFePO_4$ was prepared from the biogenic iron(II) phosphate using a hydrothermal and a calcination process. The two iron compounds had a specific morphology assembled from nanometric sheets. The biogenic iron(II) phosphate and the microbially derived $LiFePO_4$ showed electrochemical charge-discharge reactions as an anode and a cathode, respectively. The biogenic products were investigated for their potential as anode and cathode materials for lithium-ion batteries.

The mineralization processes of biogenic metal oxides have attracted much interest in terms of biology, crystal growth, and application (Ghiorse 1984; Konhauser 1998; Fortin and Langley 2005; Tebo et al. 2005; Banfield et al. 2000; Arakaki et al. 2008; Weber et al. 2006). Iron-oxidizing bacteria produce trivalent iron oxides, such as lepidocrocite (γ-FeOOH), goethite (α-FeOOH), and ferrihydrite ($5Fe_2O_3 \cdot 9H_2O$) in aqueous media (Ghiorse 1984; Cornell and Schwertmann 2003). The specific surface area of biogenic ferrihydrite was estimated to be over 90 $m^2\ g^{-1}$ (Nishina et al. 2012; Langley et al. 2009; Jambor and Dutrizac 1998). Generally, it is not easy to synthesize iron oxides with such a high specific surface area (Tuutijärvi et al. 2009; Hu et al. 2006; Jiao et al. 2006; Huang et al. 2011; Tang et al. 2011; Brezesinski et al. 2006; Tartaj et al. 2011; Zhong et al. 2007). In the second part of this chapter, one of the structural characteristics is focused: biogenic metal oxides have a high specific surface area over of 100 $m^2\ g^{-1}$ (Nishina et al. 2012; Langley et al. 2009; Jambor and Dutrizac 1998; Nelson et al. 1999; Villalobos et al. 2005). The syntheses in aqueous media at a low temperature may involve the formation of such nanostructures of biogenic iron oxides.

Recently, a microbial-mineralization-inspired approach has been shown for the syntheses of iron and manganese oxides with controlled oxidation states, crystal phases, and morphologies in aqueous solution (Oaki et al. 2012; Oaki and Imai 2007a, b; Oba et al. 2010). The selective syntheses of γ-FeOOH, α-FeOOH, and ferrihydrite were achieved in aqueous solution (Oaki et al. 2012). Since the syntheses through a microbial-mineralization-inspired approach proceed in an aqueous solution at a low temperature, the resultant oxyhydroxides could have nanostructures with a high specific surface area comparable to that of the biogenic ones. If iron oxyhydroxides with a high specific surface area are obtained as the precursors, an appropriate thermal treatment directs the formation of trivalent iron oxide (Fe_2O_3) without a decrease in the surface area. γ-FeOOH and α-FeOOH are precursors for the preparation of γ-Fe_2O_3 and α-Fe_2O_3 $m^2\ g^{-1}$, respectively. The tuned conditions for the thermal treatment led to the formation of γ-Fe_2O_3 and α-Fe_2O_3 with a high specific surface area of over 100 $m^2\ g^{-1}$. The resultant γ-Fe_2O_3 was applied to a cathode material of a lithium-ion battery and an adsorbent of an organic dye molecule (Yagita et al. 2013).

7.2 Biogenic Iron Compounds

7.2.1 Biogenic Production of Iron(II) Phosphate (Kageyama et al. 2015)

Scheme 7.1 shows the experimental procedure for the production of biogenic iron(II) phosphate and microbially derived lithium iron phosphate. The iron-reducing bacteria, *Shewanella algae* strain BrY, were obtained from the American Type Culture Collection (ATCC no. 51181). Detailed procedures for the production

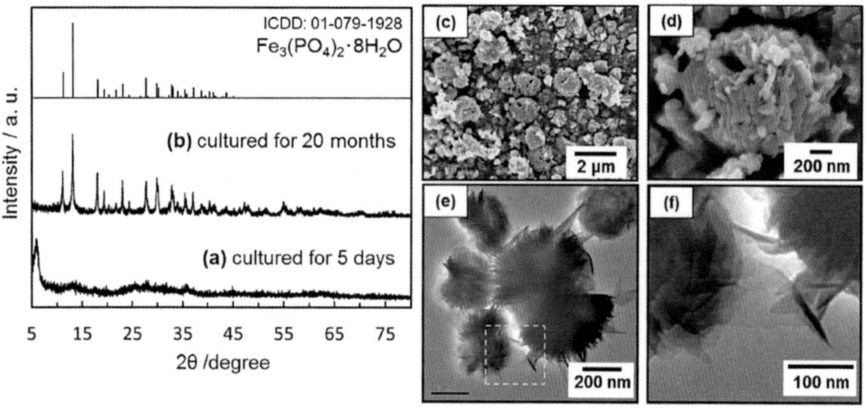

Fig. 7.1 X-ray diffraction (XRD) pattern (**a** and **b**), SEM (**c** and **d**), and TEM (**e** and **f**) images of the biogenic products cultured for 5 days. Sharp signals assignable to $Fe_3(PO_4)_2$ were observed for the products after aging in water for 20 months (**b**). The TEM image (**f**) was focused on the yellow dashed square in (**e**) (Reproduced from Kageyama et al. 2015 with permission (Copyright 2015 The Royal Society of Chemistry))

of biogenic iron(II) phosphate and subsequent reactions were described in a previous article (Kageyama et al. 2015). The microbial iron(II) phosphate sample was obtained with the reduction of iron(III) ions by a culture of the iron-reducing bacteria at 30 °C. The biogenic products were reacted with an aqueous solution dissolving $LiOH{\cdot}H_2O$ and L-(+)-ascorbic acid under a hydrothermal condition at 200 °C. The carbonization of the organic additives in the resultant products was conducted at 450 °C under an argon flow condition.

7.2.2 Characterization of Biogenic Iron(II) Phosphate

A yellow-green colored powder was obtained from the microbial medium after the culture for 5 days. The presence of a broadband around $2\theta = 6°$ suggests that the products have a layered structure (Fig. 7.1a). Scanning electron microscopy (SEM) (Fig. 7.1c, d) and transmission electron microscopy (TEM) (Fig. 7.1e, f) images showed that spherical particles 0.5–1.2 μm in diameter were composed of thin sheets ca. 50 nm in thickness. The thin nanosheets were clearly observed from the magnified TEM image (Fig. 7.1f). The specific surface area was 33.1 $m^2\ g^{-1}$. The Fe/P molar ratio was estimated to be 1.75 using inductively coupled plasma-atomic emission spectroscopy (ICP-AES). The products with low crystallinity changed into well-crystalline $Fe_3(PO_4)_2{\cdot}xH_2O$ through the aging process in water for 20 months (Fig. 7.1b). According to the weight loss in the thermogravimetry (TG) analysis, the biogenic spherical microparticles were found to be composed of $Fe_3(PO_4)_2{\cdot}xH_2O$ ($x = 4.5$) thin plates and to be accompanied by a small amount of amorphous iron compounds.

In previous works (Kashefi et al. 2001; Lovley et al. 2004; Azam and Finneran 2014), iron-reducing bacteria were reported to produce minerals on their cell

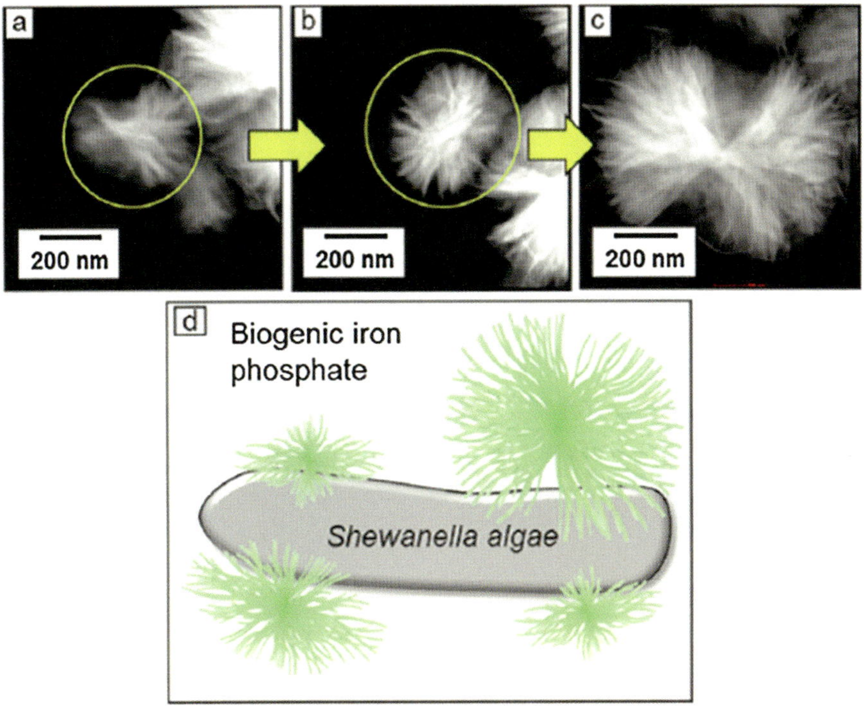

Fig. 7.2 High-angle annular dark-field scanning TEM images of biogenic iron(II) phosphate in the same batch (**a–c**). A schematic illustration of biogenic iron(II) phosphate on the iron-reducing bacteria (**d**) (Reproduced from Kageyama et al. 2015 with permission (Copyright 2015 The Royal Society of Chemistry))

surfaces. As shown in high-angle annular dark-field scanning TEM images (Fig. 7.2), the nanosheets forming a dumbbell-like structure were supposed to grow radially from the center of a spherical particle of the biogenic iron(II) phosphate on the outer membrane of the iron-reducing bacteria.

7.2.3 *Microbially Derived Lithium Iron Phosphate*

As shown in XRD patterns (Fig. 7.3a, b), the as-prepared particles after hydrothermal treatment of the biogenic products were identified as $LiFePO_4$. The crystal phase of the products was not changed by the calcination at 450 °C for the carbonization of organic additives. According to SEM and TEM observations (Fig. 7.3c–h), the morphology of the as-prepared and calcined particles was inherited from the original biogenic iron(II) phosphate. The specific surface area (33.2 $m^2\ g^{-1}$) of the calcined particles was almost the same as that of the original structure. According to the TG measurement under an air atmosphere, the calcined products contained 11.4 wt%

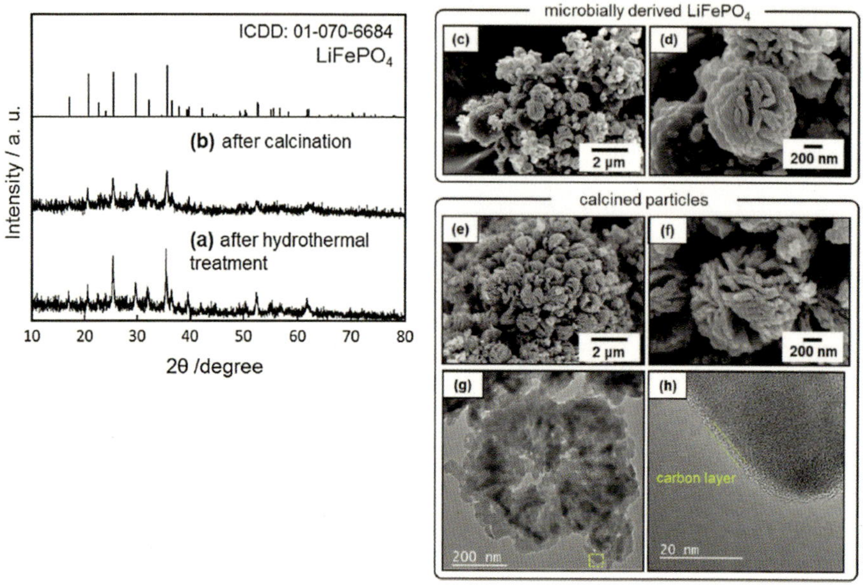

Fig. 7.3 XRD patterns (**a** and **b**) and SEM (**c–f**) and TEM (**g** and **h**) images of the microbially derived products after hydrothermal treatment (**a**, **c** and **d**) and calcined particles (**b**, **e–h**) (Reproduced from Kageyama et al. 2015 with permission (Copyright 2015 The Royal Society of Chemistry))

carbon. According to the presence of the G line at 1600 cm^{-1} in Raman spectra (Salah et al. 2006) and a layer on the surface of $LiFePO_4$ shown in the TEM image (Fig. 7.3h), the calcined products were covered with graphite-like carbon. The atomic ratio of the calcined products was estimated to be Li, 0.44; Fe, 1.60; and P, 1.00, via ICP-AES. This indicates the deficiency of lithium and the excess of iron for the stoichiometric composition of $LiFePO_4$. The excess iron existed as amorphous iron compounds.

7.2.4 Electrochemical Performance as an Anode Material

Amorphous Fe^{3+}-based oxide nanoparticles produced by *Leptothrix ochracea*, aquatic bacteria living worldwide, show potential as an Fe^{3+}/Fe^0 conversion anode material for lithium-ion batteries (Hashimoto et al. 2014). The tubular particles have a fixed bore diameter of ~1 μm, a variable length of up to several centimeters, and a surface area of 280 m^2/g. A capacity of ~600 mAh/g was maintained over at least 50 cycles. The presence of minor components, Si and P, in the original nanoparticles leads to a specific electrode architecture with Fe-based electrochemical centers embedded in a Si- and P-based amorphous matrix.

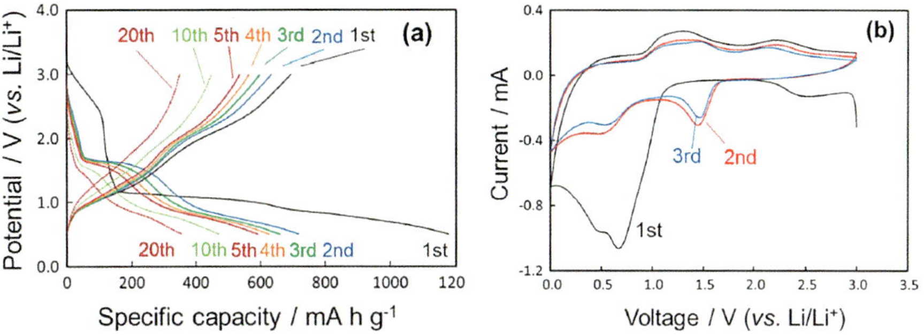

Fig. 7.4 Charge-discharge curves at 50 mM g^{-1} (**a**) and CV curves at a scan rate of 0.1 mVs^{-1} (**b**) of biogenic iron(II) phosphate (Reproduced from Kageyama et al. 2015 with permission (Copyright 2015 The Royal Society of Chemistry))

Here, the electrochemical performance of the biogenic iron phosphate is evaluated as an anode material for lithium-ion batteries. Figure 7.4a shows the charge-discharge curves with metal lithium as a counter electrode at 50 mA g^{-1}. The cyclic voltammetry (CV) curves of biogenic iron phosphate were operated at a scan rate of 0.1 mV s^{-1} (Fig. 7.4b). The open-circuit voltage was 3.4 V, and the first anodic sweep was started from 3.0 V. The first discharge capacity, 1177 mA h g^{-1}, was due to three electrochemical reactions at 2.4–3.0, 1.0–1.1, and 0.5–0.9 V. The irreversible anodic current was observed over 2.4 V and at ~0.7 V in the first anodic CV curve. The significant large capacity at the first discharge was affected by the formation of a solid electrolyte interphase (SEI) and a polymeric layer deposit (Taberna et al. 2006; Reddy et al. 2007; Das et al. 2013; Petnikota et al. 2015), the decomposition of impurities, and the reactions of polytetrafluoroethylene (PTFE) and acetylene black. The discharge capacities included 253 mA h g^{-1} (the first discharge) and ~50 mA h g^{-1} (cycles 2–20) owing to the capacities of acetylene black and PTFE. In the second discharge curve, the electrochemical reactions occurred at ~1.6 and 0.5–0.9 V. The reversible electrochemical reactions were observed at 1.45 and 0.5 V in anodic CV curves and at 1.1–1.5 and 2.2 V in cathodic CV curves. Because iron(III) did not exist in the biogenic iron(II) phosphate and the anodic current at 1.45 V was not observed in the first CV sweep, all of the initial iron(II) was considered to be oxidized to iron(III) after the first charge-discharge cycle. Thus, the electrochemical reactions at 1.45 V in the anodic curve and 2.2 V in the cathodic curve are due to the Fe^{3+}/Fe^{2+} redox couple. The reactions at 0.5 V in the anodic curve and 1.1–1.5 V in the cathodic curves are ascribed to the Fe^{2+}/Fe^{0} redox couple (Larcher et al. 2003; Cherian et al. 2012, 2013; Taberna et al. 2006; Reddy et al. 2007; Das et al. 2013; Petnikota et al. 2015). The reversible capacities in cycles 2–20 were 720–350 mA h g^{-1}. These results indicated that the electrochemical oxidation and reduction of iron were observed in the biogenic iron phosphate. The total electrochemical reaction in the first discharge was deduced as follows:

$$Fe_3(PO_4)_2 + 6Li^+ + 6e^- \rightarrow 3Fe + 2Li_3PO_4$$

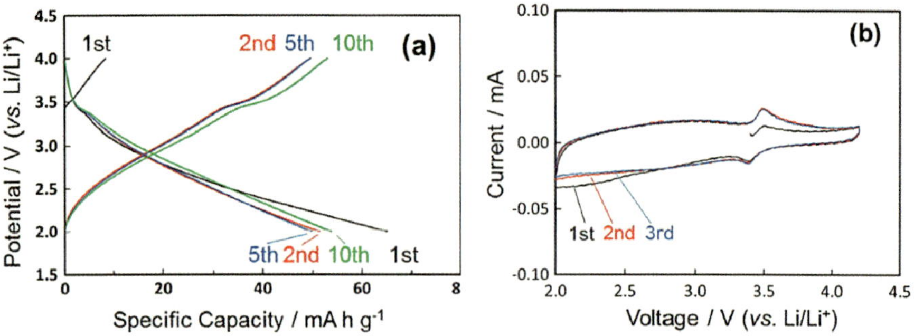

Fig. 7.5 Charge-discharge curves at 50 mM g^{-1} (**a**) and CV curves at a scan rate of 0.1 mVs^{-1} (**b**) of the calcined $LiFePO_4$ particles. The active material, acetylene black, and PTFE were mixed at a weight ratio of 50:45: 5 (Reproduced from Kageyama et al. 2015 with permission (Copyright 2015 The Royal Society of Chemistry))

The observed reversible capacities in a few cycles were larger than the theoretical capacity for the above equation (450 mA h g^{-1}). Other redox reactions are supposed to occur, owing to the amorphous iron compounds.

7.2.5 Electrochemical Performance as a Cathode Material

Figure 7.5a shows the electrochemical performance of the microbially derived $LiFePO_4$. Reversible charge and discharge reactions were clearly observed for the microbially derived electrodes. The insertion and extraction of lithium ions were observed at 3.4 and 3.5 V, respectively (Padhi et al. 1997; Yuan et al. 2011; Wang et al. 2008), as shown in the CV curves (Fig. 7.5b). Although the specific capacity of the initial charge was a very low value, the initial discharge capacity was drastically improved. Charge and discharge capacities were 40–50 mA h g^{-1} after the first cycle. The recovery of capacity suggests the electrochemical insertion of lithium ions into the lithium-deficient $LiFePO_4$ which contained iron(III). The morphology of the $LiFePO_4$ particles with a nanostructure is also deduced to be effective for the electrochemical performance. These results indicate that a practically useful cathode material can be produced from the biogenic mineral. However, the value of observed specific capacities was smaller than the theoretical specific capacity of $LiFePO_4$, 170 mA h g^{-1}. Moreover, the plateau at 3.4 V, which corresponds to the electrochemical reaction potential of $LiFePO_4$, was hardly observed in charge-discharge curves. The degradation of performance is not ascribed to low electron conductivity because the active material was covered with the graphite-like carbon coating and a high amount of acetylene black. The short plateau in charge-discharge curves is attributed to the low crystallinity of the microbially derived product and lithium defects in the $LiFePO_4$ crystals. The small capacity was due to the existence of impurities such as amorphous iron compounds.

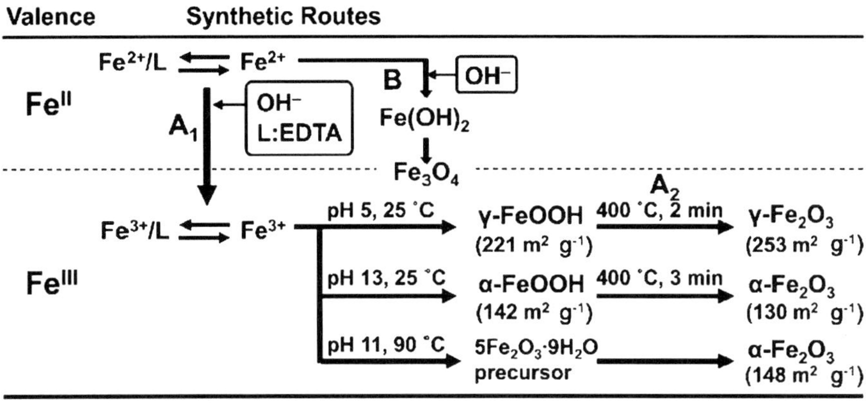

Scheme 7.2 Synthetic routes based on a microbial-mineralization-inspired approach (A_1) and the subsequent thermal treatment (A_2). Routes A and B are the syntheses with and without EDTA, respectively (Reproduced from Yagita et al. 2013 with permission (Copyright 2013 Wiley-VCH))

7.3 Bio-inspired Production of Iron Compounds

7.3.1 Syntheses of Iron Compounds (Yagita et al. 2013)

Scheme 7.2 shows the synthetic routes of iron oxides based on a microbial-mineralization-inspired approach (Yagita et al. 2013). The precursors γ-FeOOH and α-FeOOH were synthesized in an aqueous solution at a low temperature. An aqueous solution containing divalent iron(II) ions (Fe^{2+}) and disodium dihydrogen ethylenediaminetetraacetate (EDTA) as a ligand was prepared using purified water. After adjustment of the pH, the precursor solutions were maintained at 25 °C for 1 day. While magnetite (Fe_3O_4) nanocrystals were formed in the absence of EDTA through route B, the trivalent iron oxides were obtained in the presence of EDTA through route A (Scheme 7.2). Changes in EDTA concentration, pH, and temperature led to control of the crystal phases, such as α-FeOOH, γ-FeOOH, and ferrihydrite, at low temperature (Oaki et al. 2012). The resultant γ-FeOOH and α-FeOOH as precursors were treated at 400 °C for 2 and 3 min in air, respectively.

7.3.2 Characterization of Bio-inspired Iron Compounds

The precursor and the resultant materials were assigned to γ-FeOOH and γ-Fe_2O_3 on the XRD patterns, respectively (Fig. 7.6a). Based on the Scherrer equation, the size of the crystallites was estimated to be 3.5 nm for γ-FeOOH and 4.7 nm for γ-Fe_2O_3. The Fourier transform infrared (FT-IR) spectra also supported the formation of the precursor γ-FeOOH and the conversion to γ-Fe_2O_3 without any impurity phases. The resultant brown powder suggested that the crystal phase was not Fe_3O_4

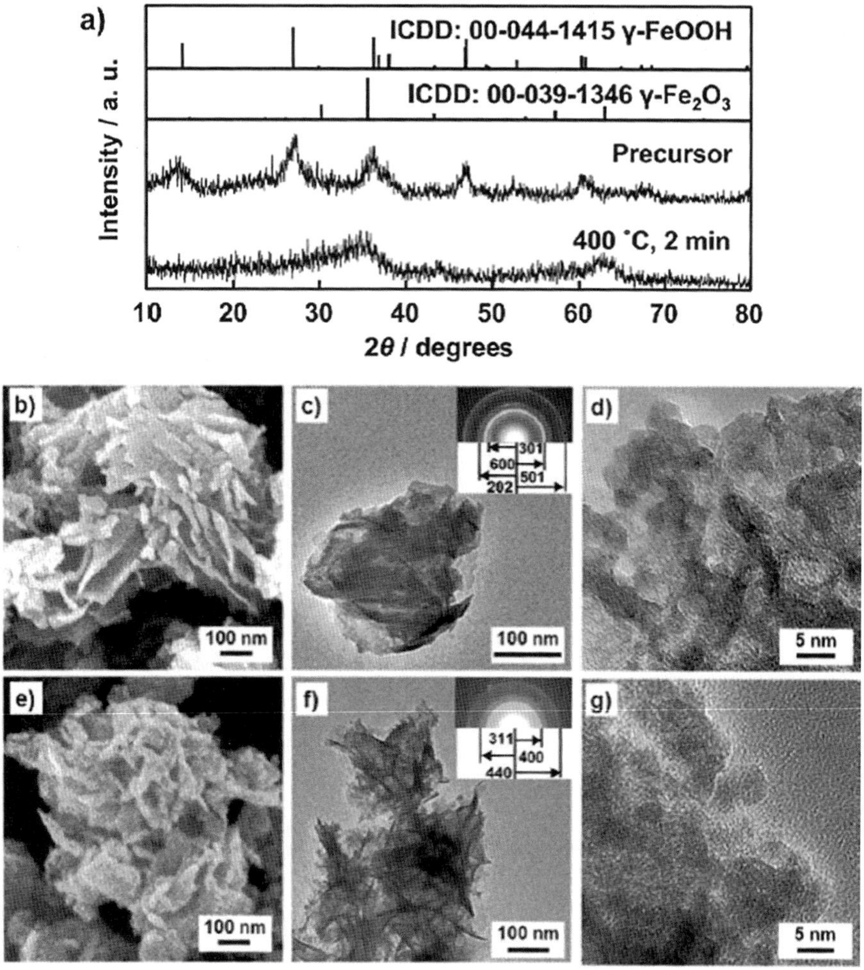

Fig. 7.6 XRD patterns (**a**) and SEM (**b** and **e**) and TEM (**c**, **d**, **f** and **g**) images with SAED patterns. (**b–d**) The precursor γ-FeOOH and (**e–g** the resultant γ-Fe_2O_3 (Reproduced from Yagita et al. 2013 with permission (Copyright 2013 Wiley-VCH))

but γ-Fe_2O_3. Based on the TG analysis, the phase transition from γ-FeOOH to γ-Fe_2O_3 was completed by the thermal treatment at 400 °C in air. Aggregates of the γ-FeOOH sheets (100–200 nm in size) were observed on the images obtained from SEM and TEM (Fig. 7.6b, c). The γ-FeOOH sheets consisted of nanocrystals less than 5 nm in size (Fig. 7.6d). The selected area electron diffraction (SAED) pattern showed Debye-Scherrer rings corresponding to the (h0L) planes of γ-FeOOH with irradiation of an electron beam in the [010] direction (Fig. 7.6c). The lattice fringes of the (301) plane were observed on the high-resolution TEM (HRTEM) image of the nanocrystals. The results suggest that the sheet-like morphology is mainly surrounded by the (010) face. However, the interior nanocrystals were not oriented in the planar direction of the (010) face. After thermal treatment at 400 °C for 2 min, the resultant γ-Fe_2O_3 showed the same hierarchical architectures as those of the precursor γ-FeOOH (Fig. 7.6e–g). Although the crystallographic orientation was not retained from the precursor γ-FeOOH (Fig. 7.6f), the lattice fringes of γ-Fe_2O_3 were clearly observed on the HRTEM image.

The BET surface areas of the precursors γ-FeOOH and γ-Fe_2O_3 were calculated to be 221 and 253 $m^2 g^{-1}$, respectively, via nitrogen adsorption isotherm. The shapes of the isotherms and the pore size distributions were not changed in either the precursor or the resultant γ-Fe_2O_3. Pores smaller than 5 nm in diameter are ascribed to the interspace of the nanocrystals. Pores ranging from 5 to 100 nm in diameter correspond to the interspace of the sheets (Fig. 7.6e, f). Based on our calculation, γ-Fe_2O_3 particles 4.9 nm in diameter are required to achieve the BET surface area of 253 $m^2 g^{-1}$. The size of the nanocrystals estimated from the BET surface area is consistent with that analyzed via TEM observations and the XRD pattern. When the time for the thermal treatment was shortened to 1 min, the precursor γ-FeOOH did not change to γ-Fe_2O_3. In contrast, the BET surface area of γ-Fe_2O_3 decreased as the time for the thermal treatment was increased to longer than 3 min. Based on the same approaches, γ-Fe_2O_3 with a BET specific surface area of 130 $m^2 g^{-1}$ was obtained from the precursor α-FeOOH (142 $m^2 g^{-1}$).

These results suggest that combination of the microbial-mineralization-inspired approach and subsequent thermal treatment has potential for the syntheses of metal oxides with a high specific surface area. A microbial-mineralization-inspired approach facilitated formation of the precursors γ-FeOOH and α-FeOOH with high specific surface area in an aqueous solution. Appropriate thermal treatment induced the phase transitions to γ-Fe_2O_3 and α-Fe_2O_3 without the collapse of the hierarchical architectures and the specific surface area. As shown in Scheme 7.2, EDTA plays an important role in the formation of trivalent iron oxyhydroxides (Oaki et al. 2012; Oaki and Imai 2007a). The complexation between EDTA and iron(II) inhibits the formation of divalent iron precipitates and the subsequent conversion to Fe_3O_4. The trivalent iron oxyhydroxides are directly obtained after the oxidation of iron(II) by oxygen dissolved in the precursor solution. Changes in the initial pH and EDTA concentration facilitate control of the crystal phases (Oaki et al. 2012). Changes in pH influence the solubility of each crystal phase and the chemical structures of the dissolved iron species. The EDTA concentration influences the activity of iron(II) related to the degree of the supersaturation of the oxyhydroxides. Therefore, γ-FeOOH and α-FeOOH were selectively synthesized through the same aqueous solution processes. The nanostructures were generated through crystal growth with the interaction of EDTA. It has been reported that γ- and α-FeOOH can be changed to γ- and α-Fe_2O_3, respectively, through topotactic conversion with dehydration reactions (Jiao et al. 2006; Huang et al. 2011; Tang et al. 2011; Brezesinski et al. 2006; Tartaj et al. 2011; Zhong et al. 2007; Cudennec and Lecerf 2005). The tuned conditions of temperature and time for the thermal treatment direct the conversion from FeOOH to Fe_2O_3 and the inhibition of grain growth by sintering. Therefore, nanostructures with a high specific surface area were obtained in the present work. Furthermore, we previously synthesized the mesocrystals of α-Fe_2O_3 through the formation of ferrihydrite in an aqueous solution at 90 °C (Scheme 7.1) (Oaki et al. 2012). Since the thermal treatment leading to grain growth was not performed in that case, the BET specific surface area of 148 $m^2\ g^{-1}$ was achieved on the α-Fe_2O_3 mesocrystal.

7.3.3 Electrochemical Performance of Bio-inspired Iron Compounds as an Anode Material

The resultant γ-Fe_2O_3 with a high specific surface area was used as a cathode material of a lithium-ion battery and an adsorbent of an organic dye. The resultant γ-Fe_2O_3 showed a higher specific capacity than that of commercial γ-Fe_2O_3 nanocrystals (Fig. 7.7). Figure 7.7a shows the charge-discharge curves of the first cycle at the current density of 0.1 C (0.024 Ag^{-1}) between the cutoff voltages of 2.0 and 4.3 V versus Li/Li^+. The resultant γ-Fe_2O_3 shows the typical charge-discharge curves of γ-Fe_2O_3 with the intercalation reaction (Hibino et al. 2007). Figure 7.7b shows the relationship between the specific capacity and the cycle number with changes in the current density. The resultant γ-Fe_2O_3 showed improved performance, such as better capacity and cycle stability, over that of a commercial γ-Fe_2O_3 nanocrystal. The resultant γ-Fe_2O_3 has a high specific surface area with broad pore size distribution of more than 5 nm. The porosity contributes to the smooth diffusion of lithium ions and an increase in the interface of the electrochemical reaction.

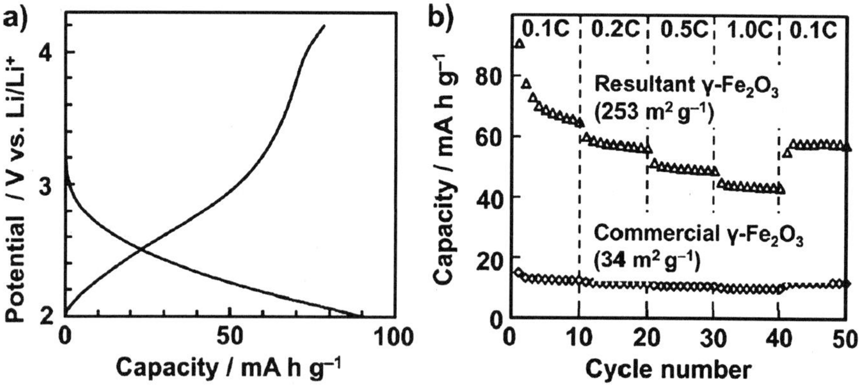

Fig. 7.7 Charge-discharge curves at the first cycle (**a**) and the relationship between the cycle number and the capacity with changes in the current density (1C = 0.024 Ag^{-1}) of the resultant γ-Fe_2O_3 (**b**) (Reproduced from Yagita et al. 2013 with permission (Copyright 2013 Wiley-VCH))

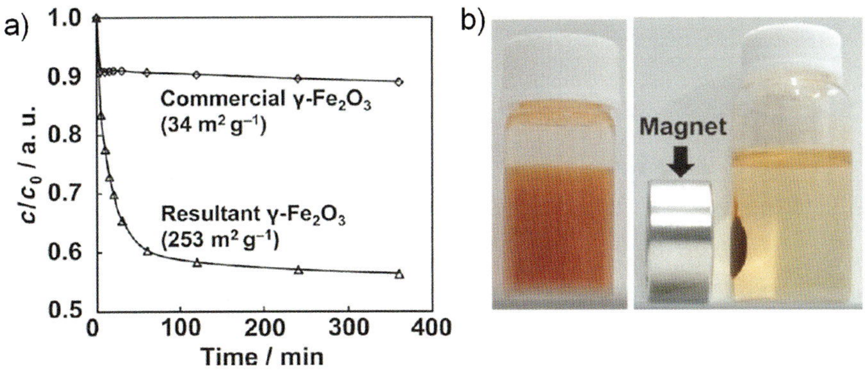

Fig. 7.8 Time-dependent adsorption behavior with the initial concentration of orange II (**a**) and γ-Fe_2O_3 adsorbed with orange II before (left) and after (right) collection with a magnet (**b**) (Reproduced from Yagita et al. 2013 with permission (Copyright 2013 Wiley-VCH))

7.3.4 Performance of Bio-inspired Iron Compounds as an Adsorbent

The resultant γ-Fe_2O_3 acted as an adsorbent for organic dye molecules (Fig. 7.8). The resultant and commercial γ-Fe_2O_3 powders were dispersed in an aqueous solution containing 0.4 mmol dm^{-3} orange II, an anionic dye, at pH 2. After adsorption for 2 h, more than 40 mol% of orange II was removed from the solution phase by the resultant γ-Fe_2O_3 powder (Fig. 7.8a). The amount of adsorbed orange II was calculated to be 0.234 g to 1.0 g of adsorbent powder. The commercial γ-Fe_2O_3 nanocrystals (50 nm in size) with a BET surface area of 34 m^2 g^{-1} adsorbed 0.07 g of orange II to 1.0 g of adsorbent powder. When the initial orange II concentration was decreased to 0.1 mmol dm^{-3}, most of the dissolved dye molecules were removed by the resultant γ-Fe_2O_3 adsorbent. After adsorption, the γ-Fe_2O_3 was easily collected by using a magnet (Fig. 7.8b). As an application of α-Fe_2O_3 with a high specific surface area, we have proposed heterogeneous catalysis of the cationic polymerization of a vinyl ether (Kanazawa et al. 2012). If the resultant α-Fe_2O_3 with the higher specific surface area is used as the catalyst, the catalytic activities will be improved.

7.4 Conclusions

We show biogenic and bio-inspired syntheses of hierarchically structured iron compounds for lithium-ion batteries. Iron-reducing bacteria produce the biogenic $Fe_3(PO_4)\cdot xH_2O(x = 1.5)$ which is applicable as an anode material for lithium-ion batteries. A cathode material for lithium-ion batteries, the microbially derived

$LiFePO_4$ was produced from the biogenic iron(II) phosphate. The biogenic iron(II) phosphate and the microbially derived $LiFePO_4$ have a unique morphology consisting of nanosheets and a high specific surface area. The two iron phosphate compounds exhibited a reversible charge-discharge reaction with lithium ions in the electrochemical system.

The precursors γ- and α-FeOOH with a high specific surface area were obtained through a microbial-mineralization-inspired approach in an aqueous solution. Subsequent thermal treatment under tuned conditions facilitated the syntheses of γ- and α-Fe_2O_3 without the collapse of the hierarchical structures and the high specific surface area. The combined methods of the microbial-mineralization-inspired approach and subsequent treatments can be applied to the synthesis of other functional transition metal oxides.

References

Arakaki A, Webb J, Matsunaga T (2003) A novel protein tightly bound to bacterial magnetic particles in Magnetospirillum magneticum strain AMB- 1. J Biol Chem 278:8745

Arakaki A, Nakazawa H, Nemoto M, Mori T, Matsunaga T (2008) Formation of magnetite by bacteria and its application. J R Soc Interface 5:977

Armand M, Tarascon J-M (2008) Building better batteries. Nature 451:652

Azam HM, Finneran KT (2014) Fe(III) reduction-mediated phosphate removal as vivianite ($Fe_3(PO_4)_2 \cdot 8H_2O$) in septic system wastewater. Chemosphere 97:1

Banfield JF, Welch SA, Zhang H, Ebert TT, Penn RL (2000) Aggregation-based crystal growth and microstructure development in natural iron oxyhydroxide biomineralization products. Science 289:751

Bernardi D, Newman J (1987) Mathematical-modeling of lithium (alloy), iron disulfide cells. J Electrochem Soc 134:1309

Brezesinski T, Groenewolt M, Antonietti M, Smarsly B (2006) Crystal-to-crystal phase transition in self-assembled mesoporous iron oxide films. Angew Chem Int Ed 45:781

Chang Y-C, Lee C-Y, Chiu H-T (2014) Porous inorganic materials from living porogens: channel-like TiO_2 from yeast-assisted sol-gel process. ACS Appl Mater Interfaces 6:31

Cherian CT, Sundaramurthy J, Kalaivani M, Ragupathy P, Kumar PS, Thavasi V, Reddy MV, Sow CH, Mhaisalkar SG, Ramakrishna S, Chowdari BVR (2012) Electrospun α-Fe_2O_3 nanorods as a stable, high capacity anode material for Li-ion batteries. J Mater Chem 22:12198

Cherian CT, Sundaramurthy J, Reddy MV, Kumar PS, Mani K, Pliszka D, Sow CH, Ramakrishna S, Chowdari BVR (2013) Morphologically robust $NiFe_2O_4$ nanofibers as high capacity Li-ion battery anode material. ACS Appl Mater Interfaces 5:9957

Cornell RM, Schwertmann U (2003) The iron oxides. Wiley-VCH, Weinheim

Cudennec Y, Lecerf A (2005) Topotactic transformations of goethite and lepidocrocite into hematite and maghemite. J Solid State Sci 7:520

Das B, Reddy MV, Chowdari BVR (2013) Li-storage of Fe_3O_4/C composite prepared by one-step carbothermal reduction method. J Alloy Compd 565:90

Ema T, Miyazaki Y, Kozuki I, Sakai T, Hashimoto H, Takada J (2011) Highly active lipase immobilized on biogenous iron oxide via an organic bridging group: the dramatic effect of the immobilization support on enzymatic function. Green Chem 13:3187

Fortin D, Langley S (2005) Formation and occurrence of biogenic iron-rich minerals Earth. Sci Rev 72:1

Gao H, Ji B, Jäger IL, Arzt E, Fratzl P (2003) Materials become insensitive to flaws at nanoscale: lessons from nature. Proc Natl Acad Sci U S A 100:5597

Ghiorse WC (1984) Biology of iron-depositing and manganese-depositing bacteria. Annu Rev Microbiol 38:515

Gorby YA, Beveridge TJ, Blakemore RP (1988) Characterization of the bacterial magnetosome membrane. J Bacteriol 170:834

Hamm CE, Merkel R, Springer O, Jurkojc P, Maier C, Prechtel K, Smetacek V (2003) Architecture and material properties of diatom shells provide effective mechanical protection. Nature 421:841

Hashimoto H, Yokoyama S, Asaoka H, Kusano Y, Ikeda Y, Seno M, Takada J, Fujii T, Nakanishi M, Murakami R (2007) Characteristics of hollow microtubes consisting of amorphous iron oxide nanoparticles produced by iron oxidizing bacteria, Leptothrix ochracea. J Magn Magn Mater 310:2405

Hashimoto H, Kobayashi G, Sakuma R, Fujii T, Hayashi N, Suzuki T, Kanno R, Takano M, Takada J (2014) Bacterial nanometric amorphous Fe- based oxide: a potential lithium-ion battery anode material. ACS Appl Mater Interfaces 6:5374

He W, Zhang X, Du X, Zhang Y, Yue Y, Shen J, Li M (2013) Bio-assisted synthesis of mesoporous $Li_3V_2(PO_4)_3$ for high performance lithium- ion batteries. Electrochim Acta 112:295

Hibino M, Terashima J, Yao T (2007) Reversible and rapid discharge-charge performance of γ-Fe_2O_3 prepared by aqueous solution method as the cathode for lithium-ion battery. J Electrochem Soc 154:A1107

Hu J, Chen GH, Lo IMC (2006) Selective removal of heavy metals from industrial wastewater using maghemite nanoparticle: performance and mechanisms. J Environ Eng 132:709

Huang X, Guan JG, Xiao ZD, Tong GX, Mou FZ, Fan XA (2011) Flower-like porous hematite nanoarchitectures achieved by complexationmediated oxidation-hydrolysis reaction. J Colloid Interface Sci 357:36

Jambor JL, Dutrizac JE (1998) Occurrence and constitution of natural and synthetic ferrihydrite, a widespread iron oxyhydroxide. Chem Rev 98:2549

Jiao F, Harrison A, Jumas JC, Chadwick AV, Kockelmann W, Bruce PG (2006) Ordered mesoporous Fe_2O_3 with crystalline walls. J Am Chem Soc 128:5468

Kageyama H, Hashimoto Y, Oaki Y, Saito S, Konishi Y, Imai H (2015) Application of biogenic iron phosphate for lithiumion batteries. RSC Adv 5:68751

Kanazawa A, Kanaoka S, Yagita N, Oaki Y, Imai H, Oda M, Arakaki A, Matsunaga T, Aoshima S (2012) Biologically synthesized or bioinspired process-derived iron oxides as catalysts for living cationic polymerization of a vinyl ether. Chem Commun 48:10904

Kashefi K, Tor JM, Nevin KP, Lovley DR (2001) Reductive precipitation of gold by dissimilatory Fe(III)-reducing bacteria and archaea. Appl Environ Microbiol 67:3275

Klaus T, Joerger R, Olsson E, Granqvist C-G (1999) Silver-based crystalline nanoparticles, microbially fabricated. Proc Natl Acad Sci U S A 96:13611

Konhauser KO (1998) Diversity of bacterial iron mineralization. Earth Sci Rev 43:91

Konishi Y, Tsukiyama T, Ohno K, Saitoh N, Nomura T, Nagamine S (2006) Intracellular recovery of gold by microbial reduction of $AuCl_4^-$ ions using the anaerobic bacterium Shewanella algae. Hydrometallurgy 81:24

Labrenz M, Druschel GK, Thomsen-Ebert T, Gilbert B, Welch SA, Kemner KM, Logan GA, Summons RE, Stasio GD, Bond PL, Lai B, Kelly SD, Banfield JF (2000) Formation of sphalerite (ZnS) deposits in natural biofilms of sulfate-reducing bacteria. Science 290:1744

Langley S, Igric P, Takahashi Y, Sakai Y, Fortin D, Hannington MD, Schwarz-Schampera U (2009) Preliminary characterization and biological reduction of putative biogenic iron oxides (BIOS) from the Tonga-Kermadec Arc, southwest Pacific Ocean. Geobiology 7:35

Larcher D, Bonnin D, Cortes R, Rivals I, Personnaz L, Tarascon J-M (2003) Combined XRD, EXAFS, and mossbauer studies of the reduction by lithium of α-Fe_2O_3 with various particle sizes. J Electrochem Soc 150:A1643

Lovley DR, Holmes DE, Nevin KP (2004) Dissimilatory Fe(III) and Mn(IV) reduction. Adv Microb Physiol 49:219

Nam KT, Kim D-W, Yoo PJ, Chiang C-Y, Meethong N, Hammond PT, Chiang Y-M, Belcher AM (2006) Virus-enabled synthesis and assembly of nanowires for lithium ion battery electrodes. Science 312:885

Nelson YM, Lion LW, Ghiorse WC, Shuler ML (1999) Production of biogenic Mn oxides by Leprothrix discophora SS-1 in a chemically defined growth medium and evaluation of their Pb adsorption characteristics. Appl Environ Microbiol 65:175

Nishina Y, Hashimoto H, Kimura N, Miyata N, Fujii T, Ohtani B, Takada J (2012) Biogenic manganese oxide: effective new catalyst for direct bromination of hydrocarbons. RSC Adv 2:6420

Oaki Y, Imai H (2007a) One-pot synthesis of manganese oxide nanosheets in aqueous solution: chelation-mediated parallel control of reaction and morphology. Angew Chem Int Ed 2007(46):4951

Oaki Y, Imai H (2007b) Chelation-mediated aqueous synthesis of metal oxyhydroxide and oxide nanostructures: combination of ligand-controlled oxidation and ligandcooperative morphogenesis. Chem Eur J 13:8564

Oaki Y, Yagita N, Imai H (2012) One-pot aqueous solution syntheses of iron oxide nanostructures with controlled crystal phases through a microbial-mineralization-inspired approach. Chem Eur J 18:110

Oba M, Oaki Y, Imai H (2010) A Microbial-mineralization-inspired aproach for synthesis of manganese oxide nanostructures with controlled oxidation states and morphologies. Adv Funct Mater 20:4279

Padhi AK, Nanjundaswamy KS, Goodenough JB (1997) Phospho-olivines as positive-electrode materials for rechargeable lithium batteries. J Electrochem Soc 144:1188

Petnikota S, Marka SK, Banerjee A, Reddy MV, Srikanth VVSS, Chowdari BVR (2015) Graphenothermal reduction synthesis of 'exfoliated graphene oxide/iron (II) oxide' composite for anode application in lithium ion batteries. J Power Sources 293:253

Reddy MV, Yu T, Sow C-H, Shen ZX, Lim CT, Rao GVS, Chowdari BVR (2007) α-Fe_2O_3 nanoflakes as an anode material for Li-ion batteries. Adv Funct Mater 17:2792

Reddy MV, GVS R, BVR C (2013) Metal oxides and oxysalts as anode materials for Li ion batteries. Chem Rev 113:5364

Salah AA, Mauger A, Zaghib K, Goodenough JB, Ravet N, Gauthier M, Gendron F, Julien CM (2006) Reduction Fe^{3+} of impurities in $LiFePO_4$ from pyrolysis of organic precursor used for carbon deposition. J Electrochem Soc 153:A1692

Shim H-W, Park S, Song HJ, Kim J-C, Jang E, Hong KS, Kim TD, Kim D-W (2015) Biomineralized multifunctional magnetite/carbon microspheres for applications in Li-ion batteries and water treatment. Chem Eur J 21:4655

Son D, Kim E, Kim T-G, Kim MG, Cho J, Park B (2004) Nanoparticle iron-phosphate anode material for Li-ion battery. Appl Phys Lett 85:5875

Taberna L, Mitra S, Poizot P, Simon P, Tarascon J-M (2006) High rate capabilities Fe_3O_4-based Cu nano-architectured electrodes for lithium-ion battery applications. Nat Mater 5:567

Tang W, Li Q, Gao S, Shang JK (2011) Arsenic (III,V) removal from aqueous solution by ultrafine α-Fe_2O_3 nanoparticles synthesized from solvent thermal method. J Hazard Mater 192:131

Tao X, Wu R, Xia Y, Huang H, Chai W, Feng T, Gan Y, Zhang W (2014a) Biotemplated fabrication of Sn@C anode materials based on the unique metal biosorption behavior of microalgae. ACS Appl Mater Interfaces 6:3696

Tao X, Zhang J, Xia Y, Huang H, Du J, Xiao H, Zhang W, Gan Y (2014b) Bio-inspired fabrication of carbon nanotiles for high performance cathode of Li-S batteries. J Mater Chem A 2:2290

Tao X, Chai W, Xu F, Luo J, Xiao H, Liang C, Gan Y, Huang H, Xia Y, Zhang W (2015) Biotemplated fabrication of highly defective carbon anchored MnO anode materials with high reversible capacity. Electrochim Acta 169:159

Tartaj P, Amarilla JM (2011) Iron oxide porous nanorods with different textural properties and surface composition: preparation, characterization and electrochemical lithium storage capabilities. J Power Sources 196:2164

Tebo BM, Johnson HA, McCarthy JK, Templeton AS (2005) Geomicrobiology of manganese(II) oxidation. Trends Microbiol 13:421

Thackeray MM, David WIF, Goodenough JB (1982) Structural characterization of the lithiated iron oxides $Li_xFe_3O_4$ and $Li_xFe_2O_3$ ($0<x<2$). Mater Res Bull 17:785

Tuutijärvi T, Lu J, Sillanpää M, Chen G (2009) As(V) adsorption on maghemite nanoparticles. J Hazard Mater 166:1415

Villalobos M, Toner B, Bargar J, Sposito G (2003) Characterization of the manganese oxide produced by pseudomonas putida strain MnB1. Geochim Cosmochim Acta 67:2649

Villalobos M, Bargar J, Sposito G (2005) Mechanisms of Pb(II) sorption on a biogenic manganese oxide. Environ Sci Technol 39:569

Wang Y, Wang Y, Hosono E, Wang K, Zhou H (2008) The design of a $LiFePO_4$/carbon nanocomposite with a core-shell structure and its synthesis by an in situ polymerization restriction method. Angew Chem Int Ed 47:7461

Weber KA, Achenbach LA, Coates JD (2006) Microorganisms pumping iron: anaerobic microbial iron oxidation and reduction. Nat Rev Microbiol 4:752

Wu B, Song H, Zhou J, Chen X (2011) Iron sulfide-embedded carbon microsphere anode material with high-rate performance for lithium-ion batteries. Chem Commun 47:8653

Xia Y, Zhang W, Huang H, Gan Y, Xiao Z, Qian L, Tao X (2011) Biotemplating of phosphate hierarchical rechargeable $LiFePO_4$/C spirulina microstructures. J Mater Chem 21:6498

Xia Y, Zhang W, Xiao Z, Huang H, Zeng H, Chen X, Chen F, Gan Y, Tao X (2012) Biotemplated fabrication of hierarchically porous NiO/C composite from lotus pollen grains for lithium-ion batteries. J Mater Chem 22:9209

Xia Y, Xiao Z, Dou X, Huang H, Lu X, Yan R, Gan Y, Zhu W, Tu J, Zhang W, Tao X (2013) Green and facile fabrication of hollow porous MnO/C microspheres from microalgaes for lithium-ion batteries. ACS Nano 7:7083

Yagita N, Oaki Y, Imai H (2013) A microbial-mineralization approach for syntheses of iron oxides with a high specific surface area. Chem Eur J 19:4419

Young JR, Davis SA, Bown PR, Mann S (1999) Coccolith ultrastructure and biomineralisation. J Struct Biol 126:195

Yuan L-X, Wang Z-H, Zhang W-X, Hu X-L, Chen J-T, Huang Y-H, Goodenough JB (2011) Development and challenges of $LiFePO_4$ cathode material for lithium-ion batteries. Energy Environ Sci 4:269

Zhong Z, Ho J, Teo J, Shen S, Gedanken A (2007) Synthesis of porous α-Fe_2O_3 nanorods and deposition of very small gold particles in the pores for catalytic oxidation of CO. Chem Mater 19:4776

Chapter 8
Biopolymer-Directed Magnetic Composites

Christian Debus, Maria Siglreitmeier, and Helmut Cölfen

Abstract The question investigated in this chapter is: Can a material obtain the advantageous material properties of multiple biominerals, when the structural elements in each model biomineral, which are responsible for these properties, are combined into one new bioinspired material? Drawing inspiration from the natural biominerals nacre, chiton teeth, and bacterial magnetosomes, a model material, containing a magnetite-gelatin composite, filling a layered scaffold extracted from natural nacre, can be synthesized.

The biopolymer gelatin has a distinct influence on the size and shape of magnetite mineralized at ambient conditions. In the gel state, gelatin can be mineralized to form superpara- and ferrimagnetic gels with tunable particle size. The ferrogel synthesis can also be transferred into demineralized nacre scaffolds, yielding layered hybrid composites.

Besides more common analytical methods, SANS is used to investigate the structure of organic and inorganic phases individually, and molecular simulations following the Kawska-Zahn approach are employed to gain insight into the earliest stages of nucleation.

8.1 Introduction

Biominerals are archetypes for advanced synthetic materials because they show complex hierarchical organic-inorganic hybrid structures with improved properties. Therefore, to copy the structural principles of biominerals in the synthesis of synthetic materials (bioinspired and biomimetic mineralization) is intensively investigated (Xu et al. 2007). The biomineral, which is probably mimicked most often, is nacre because of its extraordinary fracture toughness, which is higher than that of its main constituent aragonite (95 wt.-%) by a factor of 3000 (Currey 1977). Many layered organic-inorganic composite materials with a nacre-like structure were

C. Debus · M. Siglreitmeier · H. Cölfen (✉)
University Konstanz, Konstanz, Germany
e-mail: christian.debus@uni-konstanz.de, helmut.coelfen@uni-konstanz.de

T. Matsunaga et al. (eds.), *Biological Magnetic Materials and Applications*,
https://doi.org/10.1007/978-981-10-8069-2_8

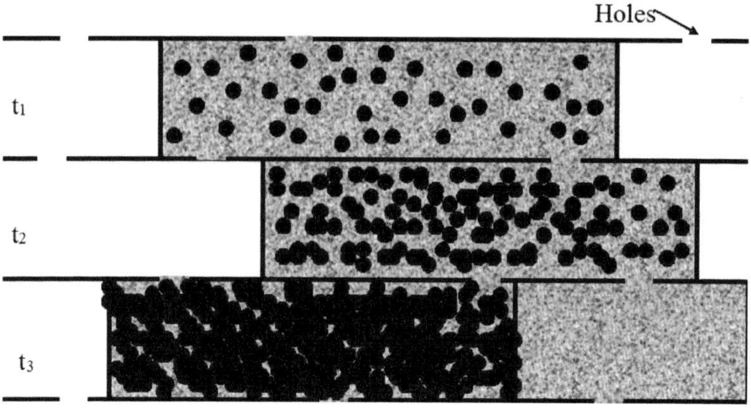

Fig. 8.1 Concept to combine the advantageous properties of nacre, chiton teeth, and magnetotactic bacteria. A demineralized nacre matrix as template with nanoholes in the chitin layers is filled with a gelatin solution, and a gelatin gel is formed upon cooling to room temperature. The gel can be infiltrated with iron ions, and subsequently magnetite is precipitated in the gel

synthesized using a variety of polymers and ceramics. Some examples like an Al_2O_3/PMMA composite material even outperformed natural nacre (Munch et al. 2008). However, irrespective of such success stories, biomimetic minerals only mimic the advantageous properties of a single biomineral. Since different biominerals have different desirable properties, it is an intriguing question if it is possible to combine the advantageous properties of several biominerals in one and the same material. Besides nacre, chiton teeth are another amazing biomineral. Chitons scratch algae from rocks, and therefore their teeth, which consist of mineralized magnetite in a protein-polysaccharide matrix, are the hardest known biomineral (Weaver et al. 2010). A third fascinating biomineral is the magnetite in magnetotactic bacteria. The magnetite nanoparticles within magnetosome vesicles are arranged along a long protein chain with their magnetic dipoles precisely aligned into a magnetically highly anisotropic structure so that they can act as a magnetic antenna for the geomagnetic field of the earth (Faivre and Schüler 2008). The magnetic field lines of the chain can nicely be visualized by electron holography to be those of a single bar magnet (Simpson et al. 2005). If the structural features of nacre, chiton teeth, and magnetotactic bacteria could be combined into a single material, the question is if such material would be simultaneously fracture resistant, extremely hard, and wear resistant and also exhibit the high magnetic anisotropy of magnetotactic bacteria. To potentially achieve these properties, demineralized nacre matrix mainly composed of chitin in a layered structure as shown in Fig. 8.1 is filled with a gelatin sol at 40 °C which yields a gelatin gel upon cooling to room temperature. This gel can then be infiltrated with iron ions in an iron solution, and subsequently, magnetite can be formed by the coprecipitation method or the partial oxidation method (Helminger et al. 2014). This mimics the aragonite formation inside a silk hydrogel in nacre or the magnetite formation in a protein-polysaccharide gel in chiton teeth.

Due to the nacre-chitin matrix, the resulting hybrid material has the structure of nacre with the magnetite gel moieties mimicking the magnetite in chiton teeth. If the magnetite nanoparticles can be aligned during the synthesis, a coupling of the magnetic dipoles can potentially be achieved to mimic the chain of aligned magnetic dipoles in magnetotactic bacteria. This book chapter describes synthesis pathways to such material.

8.2 Synthesis of Magnetite Nanoparticles in Aqueous Solution at Ambient Conditions

Magnetite nanoparticles can easily be synthesized free of additives and at ambient conditions by coprecipitation of stoichiometric amounts of iron (II) and iron (III) salts (Kang et al. 1996) or by partial oxidation of iron (II) salts (Sugimoto and Matijević 1980), yielding particles in two distinct size ranges. The particle size, shape, and composition depend strongly on the reaction conditions but can also be influenced by the presence of soluble biopolymers, as shall be elucidated later.

8.2.1 Coprecipitation

Injecting a solution of iron (III) and (II) salts in a ratio of 2:1 into a basic solution yields magnetite nanoparticles in a typical size range of 5–20 nm. In previous studies, it was shown how factors like pH, ionic strength, counterions, and reaction temperature have a distinct influence on particle morphology and composition (Vayssières et al. 1998; Jolivet et al. 1996; Taylor et al. 1987). The particles are crystalline with a typical saturation magnetization of about 30–50 emu/g (Sun et al. 2007). In this size regime, superparamagnetic behavior is observed at room temperature.

These particles can be prepared by injection of an aqueous $FeCl_{2/3}$ solution into a strongly basic NaOH solution. A black precipitate is formed that can be magnetically attracted and is washed with Millipore water to yield the pure magnetite nanoparticles. As magnetite has its point of zero charge at pH = 6.5 (Tewari and McLean 1972), the particles are highly aggregated. Particle aggregation can be avoided by the interaction of biopolymer additives and highly charged magnetite nanoparticles. Therefore, this will not pose a problem for the synthesis of ferrogels, as in these gels, the particles are attached to the organic gel.

The average particle size for the approach presented here was determined by TEM measurement to be 12 nm (Fig. 8.2a). While it is difficult to differentiate magnetite and maghemite with diffraction methods such as SAED and XRD, both phases can easily be differentiated by their different IR spectra (Fig. 8.2b) (Cornell and Schwertmann 2003). Thus, it could be shown that the iron oxide phase of the particles was in fact mainly magnetite.

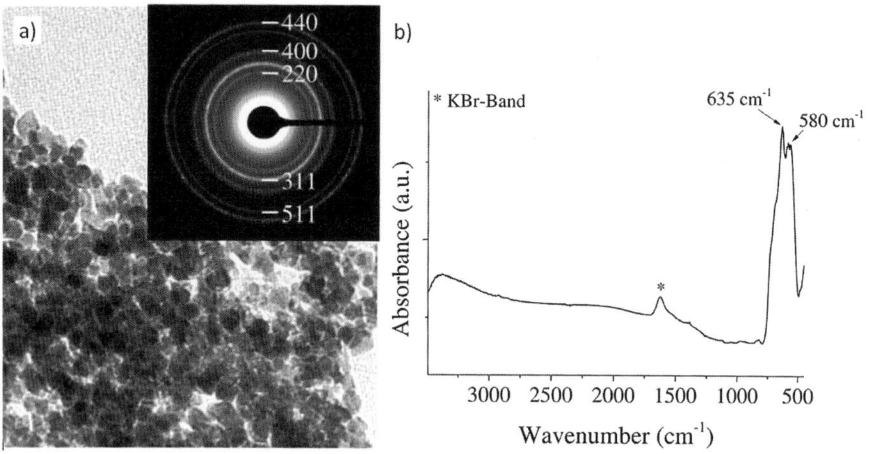

Fig. 8.2 (**a**) TEM image and SAED inset of the magnetite nanoparticles prepared by the coprecipitation method, (**b**) IR spectrum of a representative particle sample indicating the presence of magnetite

8.2.2 Partial Oxidation

Magnetite particles with a size of 50–90 nm can be synthesized by addition of an iron (II) sulfate solution to a solution of potassium hydroxide and potassium nitrate. First, a fluffy, gray-green iron (II) hydroxide gel is formed that is then gradually oxidized to magnetite in the presence of nitrate (Sugimoto and Matijević 1980). With an excess of Fe^{2+}, big spherical particles with sizes up to more than a micrometer can be formed (Vereda et al. 2013). The formation mechanism of these particles is discussed to include the formation of nanometer-sized precursor particles. The Fe^{2+} excess keeps the pH close to the point of zero charge of magnetite, so that these particles aggregate and recrystallize, forming spheres. With a controlled excess of OH^-, well-defined, octahedrally shaped particles can be formed. At alkaline pH, the primary particles are stabilized by their surface charge and can grow instead of aggregating. While many studies on the aging of iron (II) hydroxide gels were carried out at elevated temperatures (Vereda et al. 2008, 2013), the work of Siglreitmeier et al. underlined the possibility to carry out the reaction at room temperature, yielding octahedral particles with an average size of approx. 70 nm (Siglreitmeier 2015). These particles show ferrimagnetic behavior at room temperature.

8.3 Influence of Soluble Biopolymers

Biopolymers can have a distinct influence on the nucleation and morphology of nanoparticles.

One cheap and easily obtainable biopolymer with well-explored properties is gelatin, which is a partial hydrolysate of the collagen found in connective tissues from, e.g., the skin, bone, and tendons. At room temperature, gelatin is completely soluble below a concentration of approx. 1 wt-% (depending on gelatin type and quality). It does reversibly form hydrogels at temperatures around 35–40 °C (Wandrey et al. 2010).

Gelatin is widely used in food and medical applications for its ability to form hydrogels (Reinhard Schrieber 2007). Collagen, the natural structural protein from which gelatin is derived, has a rather complex, hierarchical structure. Its basic building blocks are α-helical chains consisting of a $(Gly\text{-}X\text{-}Y)_n$ repetition pattern, where X is mostly proline and Y is mostly hydroxyproline (Brodsky and Werkmeister 2004). These left-handed α-chains form a right-handed triple helix, also called tropocollagen. Most forms of collagen form fibrils that consist of aligned triple helices. These structures are stabilized by hydrogen bonds and covalent cross-linking, e.g., via oxidative cross-linking of lysine.

Gelatin as a partial hydrolysate of collagen consists of polydisperse peptide fragments that can partially re-naturate into triple helical structures. These short triple helical segments act as cross-linking points to form a three-dimensional meshed network.

Simulation studies on the interaction between iron ions and collagen give an important impression on the earliest stages of iron (hydr)oxide nucleation in gelatin hydrogels. These molecular simulations were based on work on the more well-known collagen – calcium phosphate system and system parameters were chosen accordingly to allow for good comparability (Tlatlik et al. 2006). Iron species can be complexed by the hydroxyl and carboxy functionalities of the collagen (see Figs. 8.3 and 8.12). A notable disturbance of the collagen structure could not be

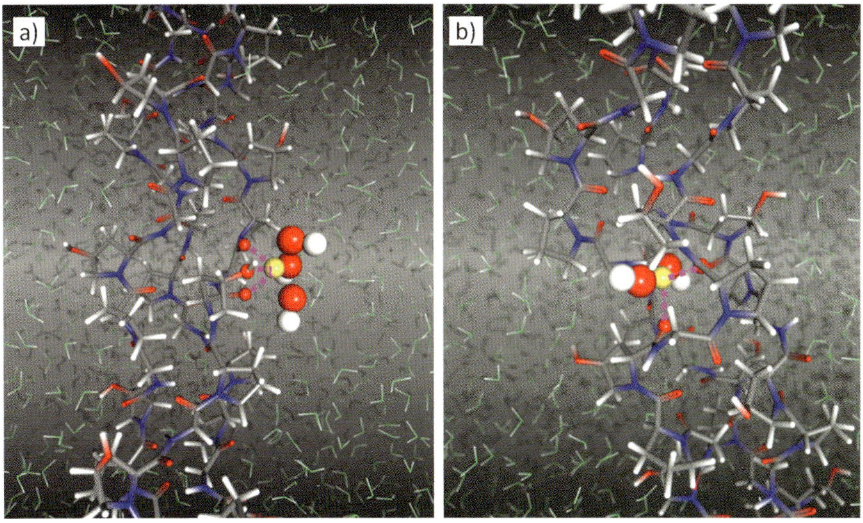

Fig. 8.3 Representative structures for the coordination of (**a**) Fe^{3+} and (**b**) Fe^{2+} to the collagen triple helix. Both ions are coordinated by multiple carbonyl and hydroxyl functional groups of the collagen (The figure was reproduced from Helminger et al. 2014)

observed. The iron (III) ion is octahedrally surrounded by oxygen atoms from the collagen functional groups and additional hydroxide ions from the solution. These observations lead to the conclusion that collagen and gelatin may be suitable nucleation sites for magnetite and lead to a very early attachment of the growing particles to the gelatin hydrogel as well as incorporation of gelatin into the particles. Interestingly, particle orientation along the gelatin fibers was in fact observed experimentally (Helminger et al. 2014). Simulation work in this direction is still ongoing, now also investigating later stages in nucleation like the formation of magnetite motifs and interface formation/ripening processes.

To understand the formation process of both ferrogels and the more complex layered materials based on them, which will be presented later in the chapter, the influence of gelatin on the particle formation and morphology shall first be explained. Gelatin degrades notably at elevated temperatures above room temperature (Pouradier and Venet 1952); thus, the influence of gelatin was examined at room temperature. Concentrations below the gelation concentration of around 1 wt-% at room temperature (Wandrey et al. 2010) were chosen.

With particles created by the coprecipitation method, a distinct influence of the biopolymer concentration on the particle size is observable. To observe this influence, the preparation method was modified from Baumgartner et al. (Fig. 8.4) (Baumgartner et al. 2013a). In an inert atmosphere, a solution of gelatin was set to the desired pH, and the pH was kept constant to only observe the influence of the biopolymer. Then, a solution containing stoichiometric amounts of iron (II) and (III) chloride was added at a defined flow rate while monitoring the pH and keeping it constant by injection of additional base.

At pH = 9, the influence of gelatin inhibiting the nucleation of magnetite is rather obvious. If the flow rate is low enough,[1] no crystalline magnetite is observed at all. Amorphous clusters with sizes averaging around 3–4 nm can be observed for the examined gelatin concentrations between 0.1 and 1 mg/mL, compared to the average particle size of 19 nm (Fig. 8.5a) that is reported for the additive-free system (Baumgartner et al. 2013a). At a higher flow rate,[2] compared to the additive-free system, the lowest gelatin concentration examined (0.1 mg/mL) leads to a decrease of the average particle size to 12 nm. These particles are still crystalline magnetite as observed by SAED and synchrotron XRD. Gelatin concentrations up to 10 mg/mL lead to a consecutive decrease of the particle size. At this gelatin concentration, the particle size is lowered to an average size of around 3 nm, which no longer shows a crystalline SAED pattern (see Fig. 8.5c) and shows very broad XRD patterns indicating the possible presence of ferrihydrite, which is discussed as a precursor phase in magnetite formation (Baumgartner et al. 2013b). At an increased pH of 11, the inhibitory effect of the gelatin seems to be attenuated, as results comparable to the experiments at pH = 9 are already found at lower flow rates. Again, a decrease in particle size is observed for increased gelatin concentrations. The particles with

[1] 1 μL/min of 1 M $FeCl_{2/3}$ solution into 10 mL gelatin solution with set pH.

[2] 10 mL/min of 1 M $FeCl_{2/3}$ solution into 10 mL gelatin solution with set pH.

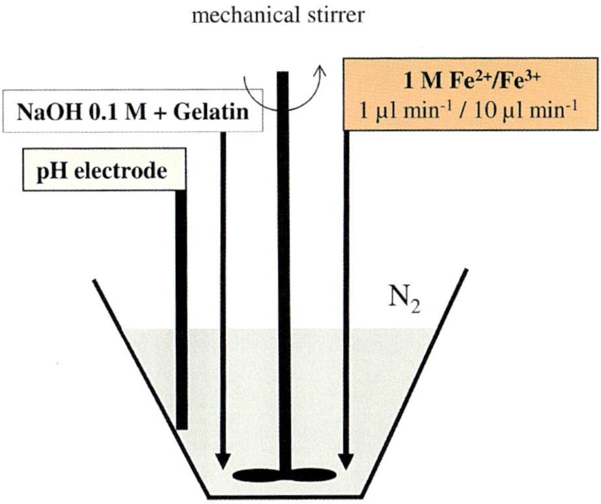

Fig. 8.4 Schematic titration setup

sizes around 3–4 nm are no longer showing crystalline diffraction patterns. In general, only minor impurities besides the desired magnetite were observed. It is notable that at this pH, in all samples with a high flow rate or low gelatin concentration, the particles possess a rather rough surface profile and there is a second, smaller particle subspecies which might again indicate the presence of a poorly ordered precursor species, able to agglomerate, fuse, and crystallize into magnetite (Baumgartner et al. 2013b).

Magnetite particles created by partial oxidation of $FeSO_4$ show a more complex change in size and morphology with increasing gelatin concentration. At a sufficiently low gelatin concentration of 0.1 mg/mL, octahedral particles are formed that still look very much like the particles synthesized without gelatin, albeit with more rounded edges and inhomogeneities in particle density, as suggested by TEM. Again, the particle size decreases when compared to an additive-free synthesis.

At a higher gelatin concentration of 1 mg/mL, the particles develop a more flowerlike spherical structure (Fig. 8.5d). Surprisingly, the average particle size increases compared to the additive-free particles. The presence of particles in the 5 nm size range as a by-product suggests that the flowerlike particles may be formed by agglomeration of these nanoparticle subunits, following a nonclassical crystallization pathway (Niederberger and Colfen 2006). Detailed TEM images and the rough, granular surface observed in SEM support this suggestion (Fig. 8.6). SAED of single particles showed single-crystalline diffraction patterns. The particle shape might suggest a formation mechanism by aggregation of sub-particles as it is also observed for other iron oxide systems (Penn and Banfield 1998; Penn and Soltis 2014).

While thermogravimetric analysis of the particles did not show a measurable amount of an included organic phase inside the magnetite nanoparticles, small organic inclusions are often found in the similar calcium carbonate and phosphate-

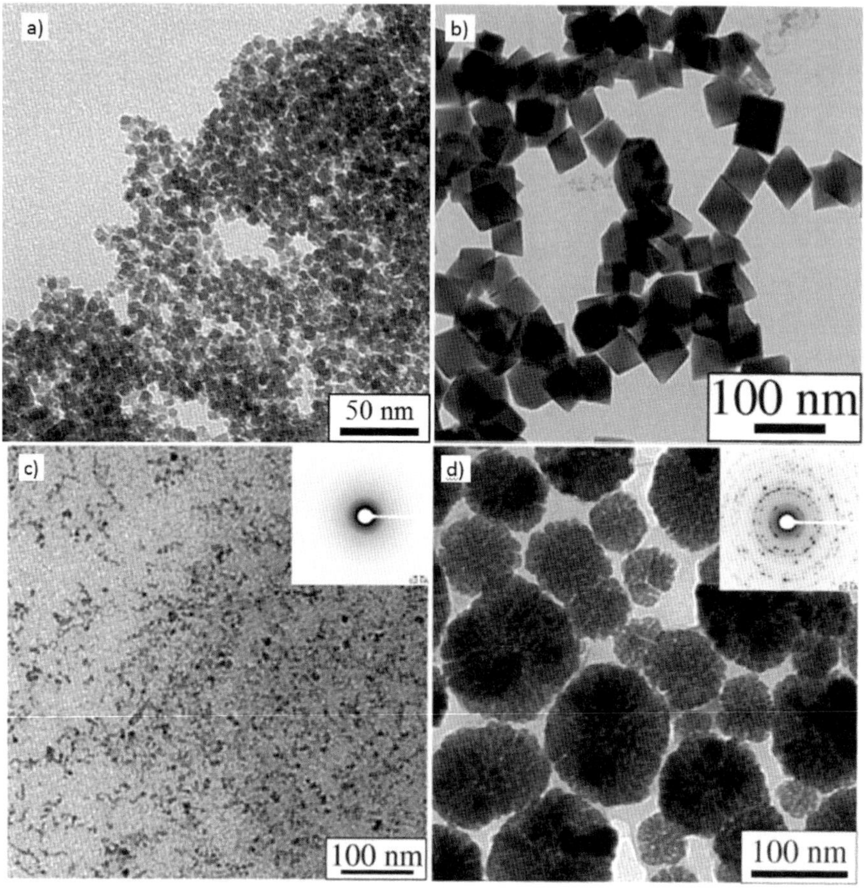

Fig. 8.5 TEM images of particles synthesized by the coprecipitation method (**a**, **c**) and the partial oxidation method (**b**, **d**); upper row without additives, lower row with gelatin concentrations of 1 mg/mL

gelatin systems (Li et al. 2009; Asenath-Smith et al. 2012; Kniep and Simon 2007) and cannot be completely excluded here either.

Examples for further peptides aiding in the nucleation and manipulating the particle morphology of magnetite in analogy to the magnetite nucleator proteins are seen during bacterial magnetosome formation (Arakaki et al. 2003). The influence of the C-terminal segment of the bacterial peptide mms6 as well as other short peptide segments, which showed magnetite-binding properties in phage displays, was investigated. The peptides have a merely marginal influence on the particles synthesized by the coprecipitation method. Looking at the partial oxidation method, under the chosen conditions, the mms6 segment also does not influence the particle formation. Short peptides chosen in phage displays for their magnetite-binding capacity led to a distinct increase in particle size while not further influencing the particle morphology.

Fig. 8.6 SEM images of particles synthesized by the partial oxidation method with a gelatin concentration of 1 mg/mL showing the rough, granular particle surface

8.3.1 Magnetic Properties

Bulk magnetite is a ferrimagnetic material. When decreasing the grain size of magnetite, finite-size- and surface effects become obvious (Teja and Koh 2009). While a macroscopic grain contains multiple magnetic domains, below a certain particle size, the formation of domain walls is energetically unfavorable, and stable single domain (SSD) particles are observed. Since magnetization reversal can no longer take place by displacing domain walls, particles with sizes at the upper end of the single domain regime show the highest coercivity. Decreasing the particle size below a value of approx. 20 nm leads to vanishing of the coercivity and of hysteresis in M(H) curves at room temperature. Due to thermal fluctuations, the particles can no longer hold a magnetic moment. Lowering the temperature below the blocking temperature T_B decreases these fluctuations far enough that the particle is "blocked," again showing coercivity and hysteresis.

SQUID measurements allow for measuring both the field and temperature dependence of a sample's magnetic moment. For the particles synthesized by the coprecipitation method, saturation magnetization of M_s = 40 emu/g can be measured (Fig. 8.7a), which is lower than the value for bulk magnetite due to the drastically increased surface area and the effect of spin disorder on the particle surface (Goya et al. 2003). The M(H) curve at room temperature shows no hysteresis. Furthermore, the zero-field-cooled M(T) measurement shows a peak indicating the blocking temperature at T_B = 140 K (Fig. 8.7b). Thus, the particles show superparamagnetic behavior as would be expected for their average size of 12 nm.

The particles synthesized by the partial oxidation method show ferrimagnetic behavior. Hysteresis is observed at room temperature, and saturation magnetizations of 74 emu/g at room temperature and 83 emu/g at 2 K are reached (Fig. 8.7c). This is comparable to similar synthetic approaches in literature (Vereda et al. 2008;

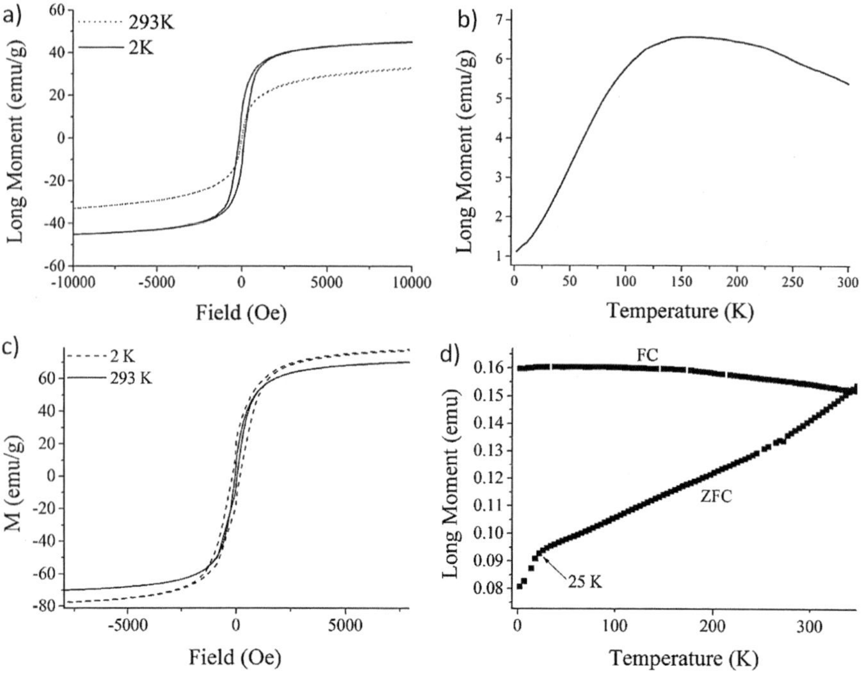

Fig. 8.7 M(H) curves and M(T) curves recorded of sample prepared by the coprecipitation method free of additives (**a**, **b**) and by the partial oxidation method with 1 mg/mL gelatin (**c**, **d**)

Vergés et al. 2008). No blocking temperature can be observed, but a bend in the zero-field-cooled M(T) curve indicates the Verwey transition temperature (Fig. 8.7d).[3] While observed at 110–120 K in bulk magnetite, it is known that this value can be pushed to lower values by small particle sizes and deviations in stoichiometry (Özdemir et al. 1993).

The flowerlike particles synthesized in the presence of 1 mg/mL of gelatin do also show ferrimagnetic behavior with only very minor changes in saturation magnetization and Verwey transition temperature. These deviations can be attributed to the slight changes in average particle size and the overall different morphology.

8.3.2 Conclusion

Aqueous syntheses for both superparamagnetic and ferrimagnetic magnetite nanoparticles under ambient conditions were presented. By changing the synthesis method and iron concentration, particle size spanning almost an order of magnitude,

[3]The Verwey transition is a crystallographic transition observed in magnetite, associated with changes in remanence, coercivity, and susceptibility (Walz 2002).

ranging from approx. 12 to 90 nm, can be realized. Their magnetic properties vary depending on their size from superparamagnetic to ferrimagnetic. Depending on reaction conditions and concentrations, gelatin as a freely available polymer can have a distinct influence on the particle size and morphology when added as a soluble additive. Molecular simulations show how collagen can pose as a possible nucleation site for both presented synthetic approaches. An increase in gelatin concentration leads to a decrease in particle size for particles prepared by the coprecipitation method. For the partial oxidation method, a change in particle morphology is also observed for increased gelatin concentrations. Particles possess an inhomogeneous, flowerlike appearance, suggesting growth by agglomeration of subunits, but still show single-crystalline behavior.

8.4 Magnetic Hydrogels

There are two main ways of commercial gelatin extraction. Gelatin is either extracted after acidic pretreatment (type A gelatin) of bovine hide or after basic pretreatment (type B gelatin) from the cattle hide and bone. One of the many differences between these two gelatin types is their stability in acidic and basic conditions. While type A gelatin is more stable in acidic conditions, type B is less prone to hydrolysis in basic conditions (Ames 1947). Since the precipitation reaction to mineralize the ferrogels takes place at pH = 13, type B gelatin is preferably used.

Gelatin dissolves/melts around 40–50 °C and will settle into a gel after cooling at concentrations above approx. 3 wt-%, depending on the gelatin grade. This makes it easy to cast gelatin gels into any desired shape.

8.4.1 Superparamagnetic Ferrogels

The synthesis protocols for ferrogels are adapted from the synthesis of magnetite nanoparticles in solutions presented before (Fig. 8.8) (Helminger et al. 2014). To synthesize superparamagnetic ferrogels, a gelatin gel is loaded with iron salts by placement into a stoichiometric $FeCl_{2/3}$ solution with a ratio of Fe (II)/Fe (III) of 1:2, turning the gel from slightly yellow to an intense orange color. The strong interaction between iron ions and gelatin that is suggested by molecular simulations is also observed when gelatin gels are impregnated with iron ions, whereas iron (III) will act as a cross-linker, reinforcing the gel, while iron (II) will have a substantially weaker influence on the hydrogel's mechanical properties.

Magnetite nanoparticles can then be precipitated within the gel via placement into a sodium hydroxide solution. This yields ferrogels with an inorganic content of approx. 20–30 wt-%. Repetitions of this mineralization cycle allow for an inorganic content of up to 70 wt-% in the dried state.

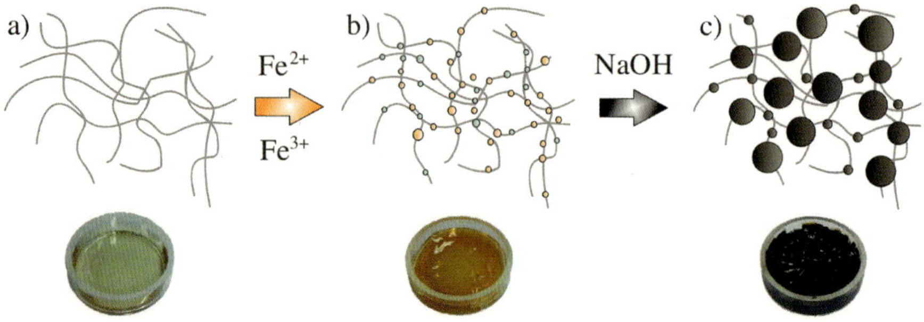

Fig. 8.8 Schematic representation of the ferrogel synthesis. (**a**) Plain gelatin hydrogel, (**b**) hydrogel loaded with ferrous and ferric chloride, (**c**) ferrogel with homogeneously distributed magnetite nanoparticles after in situ coprecipitation (The figure was reproduced from Helminger et al. 2014)

After the first mineralization cycle, the ferrogels can still be heated up and molded into a different shape, resettling into a solid gel upon cooling. Two pieces of ferrogels can also be attached to each other at slightly elevated temperatures (30 °C), thus showing self-healing properties.

Microtome cuts of these ferrogels show average particle sizes of approx. 10 nm irrespective of the gelatin concentration of the original gel (Fig. 8.9a). It shall be noted here that this matches well with the nanograin size of 11 nm which was observed for the aragonite platelets in nacre via scattering experiments (Siglreitmeier et al. 2015). The particles are spherical and thus identical in appearance to the particles synthesized in water. EDX shows a homogeneous distribution of magnetite on the micrometer scale in the gels. In swollen, embedded ferrogels, an orientation along the gelatin fibers can be observed, showing the strong protein interaction (Fig. 8.9a). This also expresses itself in the very low degree of aggregation, especially when compared to the synthesis in solution. Furthermore, the swelling ability of the ferrogel is decreasing with increasing mineral load, suggesting that the particles may act as cross-linkers between gelatin strains.

Using small-angle neutron scattering (SANS) and ultrasmall-angle neutron scattering (USANS), the organic/inorganic hybrid structure of biological and biomimetic materials can be investigated in situ in a nondestructive way.

The investigation of the gelatin hydrogels using SANS and USANS revealed the existence of two level structures in the hydrogels: one is the colloid-like large clusters filled mainly with densely packed gelatin triple helices; and the other is a 3D cage-like gel network composed of gelatin triple helices and random coils. In the case of 18 wt.-% gelatin gels, SANS/USANS results ($Q < 0.2\ nm^{-1}$) suggest that the large clusters/inhomogeneities have a size of around 400 nm (assuming a diameter for spherical shape, Roe 2000) and take up large-scale mass fractal networks. These are colloidal structures inside the gelatin that consist of renatured collagen fractions (Wu 2015). In the Q regime between 0.2 and 1.5 nm^{-1}, a power law exponent of 1.2 supposedly originates from the rodlike bundles formed by gelatin triple helices. The average 3D cage size of the gelatin hydrogel can be estimated from the transition between the two different regimes of $Q^{-\alpha}$ (Q_c ~0.24 nm^{-1}) with an estimated 3D cage

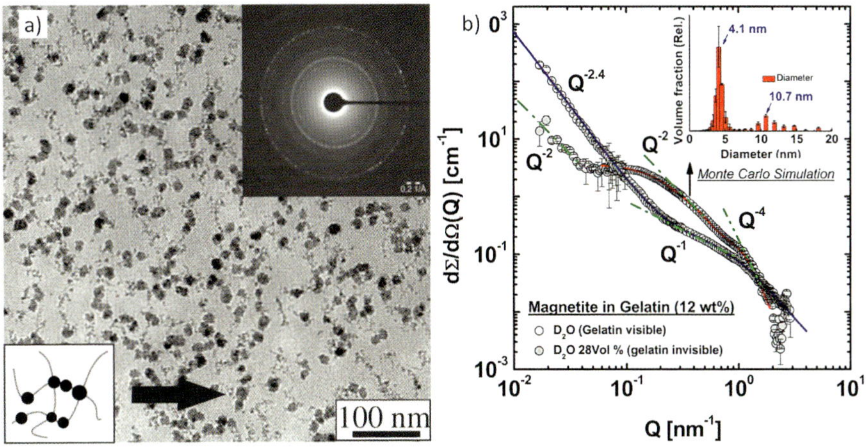

Fig. 8.9 (**a**) TEM image of an ultramicrotome cut of an embedded ferrogel after six mineralization cycles (The figure was reproduced from Helminger et al. 2014). (**b**) SANS measurements of a ferrogel with either the organic or the inorganic components hidden by contrast matching

size of approx. 26 nm. The investigation of different gel concentrations shows an increase of triple helical bundle structures with increasing gel concentration but only a minor influence on the hydrogel 3D cage size.

Contrast variation is a valuable tool to investigate multicomponent systems via SANS, which allowed to independently explore the inorganic mineral structure as well as the organic component in biominerals and biomimetic materials (Heiss et al. 2007; Pipich et al. 2008). The scattering length density of the solvent can be adjusted through contrast variation using a varying ratio of H_2O/D_2O to highlight various parts of the hybrid structure. Thus, it allows for hiding selective components of a multicomponent material by repetition of the SANS measurement with differing H_2O/D_2O ratios. In the case of the magnetite/gelatin system, this method allows for the investigation of only the gelatin component when measuring in pure D_2O and of the magnetite component in 28% D_2O. In a typical SANS contrast variation measurement (Fig. 8.9b), SANS scattering of the gelatin in ferrogels (D_2O) shows hardly any change in the gel structure (3D cage-like) besides an increase in triple helical content which probably originates from interaction with the $Fe^{2+/3+}$ ions. In 28 Vol% of D_2O, the measurement which matches the gelatin scattering and visualizes the magnetite nanoparticles, the magnetite particle size distribution obtained by a Monte Carlo simulation on the scattering curve appears consistent with the TEM measurements (larger particle sizes of ~10.7 nm coexist with smaller particles of ~4.1 nm). The magnetite scattering also reveals slight correlation between the magnetite particles which is even more pronounced in small-angle X-ray scattering experiments.

The SAXS results suggest that the gelatin concentration of the ferrogels has little influence on the magnetite particle size. In the dried samples, a correlation peak is visible, showing an average distance of the particles that is close to the particle size, thus suggesting tight packing of the magnetite particles in the dried state.

As is to be expected for this particle size, the ferrogels show typical superparamagnetic behavior when analyzed by SQUID. Hysteresis is observed below the blocking temperature and is absent at room temperature. The blocking temperature, as determined by a peak in the ZFC curve, is around 130 K and shows a tendency of rising with lowered gelatin concentration (with the magnetite content staying the same). This is explained by denser particle packing in the dried state and thus stronger dipolar interactions between the particles. The measurements give saturation magnetizations of around 26–28 emu/g. Embedding in a ferrogel does thus not change the magnetic properties of the particles notably besides lowering the saturation magnetization.

Since the particles are bound to the gelatin gel via electrostatic interaction, possible applications of these gels such as biocompatible external field-driven actuators are suggested (Siglreitmeier 2015).

8.4.2 Ferrimagnetic Ferrogels

To generate ordered ferrogels with anisotropic magnetic properties as inspired by the magnetosomes of magnetotactic bacteria, the particle size needs to be increased. The synthesis described in Sect. 8.2.2 is therefore adapted to suit the mineralization of gelatin gels. Similar to the coprecipitation method described before, a gelatin sample is placed in a $FeSO_4$ solution and then placed in a solution of KOH as a base and KNO_3 acting as a mild oxidizing agent. In accordance with the observations in solution, the piece of gelatin first turns blue, indicating the formation of the ferrous hydroxide. Consecutive repetitions of this mineralization cycle lead to a blackening of the sample, indicating the formation of magnetite nanoparticles. In thick samples, a zone of blue-gray hydroxide can still be observed in cross sections and is moving deeper into the sample with each mineralization cycle. Variation of the iron concentration between 0.2 and 0.3 M leads to a particle size between 45 and 90 nm as determined by TEM. This is interesting as this particle size is bigger than the average mesh size of the gelatin gel, which was determined by SANS to be around 22 nm. Either gelatin is incorporated into the magnetite particles, or the particles widen the gelatin mesh.

Since the XRD diffraction patterns of magnetite and maghemite are very similar, synchrotron radiation was used to investigate low-intensity peaks, which proved that the main mineral phase present in the particles is in fact magnetite. Minor iron oxide impurities were also often found. Particle size analysis by evaluation of peak broadening by the Scherrer equation showed no increase in the average particle size between multiple reaction cycles.

As is expected for particles in this size range, ferrogels with both particle sizes show typical ferrimagnetic behavior with a saturation magnetization of 74 emu/g at

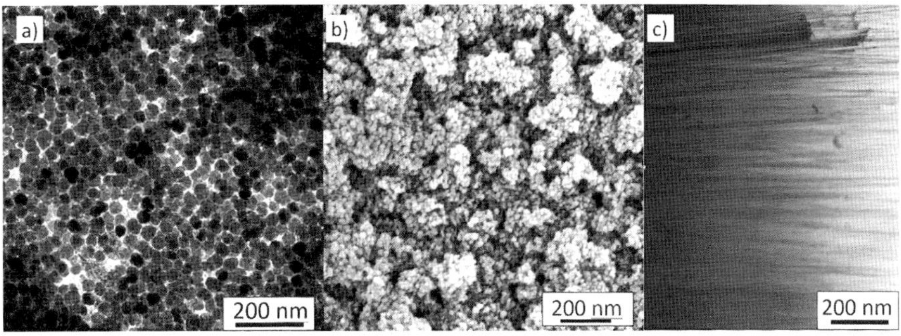

Fig. 8.10 (**a**) Ultramicrotome cut from a sample prepared with 0.3 M $FeSO_4$ after four reaction cycles, (**b**) SEM image of a dried ferrogel prepared with 0.2 M $FeSO_4$ after four reaction cycles, (**c**) light microscopic image of a ferrogel dried in a magnetic field, showing the orientation of the gel into macroscopic fibers which are aligned with the direction of the applied field

2 K and 67 emu/g at room temperature. Hysteresis is observed for the whole temperature range, and ZFC/FC curves show no indication of a blocking temperature.

Due to the ferrimagnetic nature of the particles, anisotropic gels can be achieved when drying inside a magnetic field. The formation of macroscopic fibers can be observed by light microscopy (Fig. 8.10c). Although on a different length scale, these fibers resemble the fibers observed in chiton teeth. This orientability also suggests anisotropic magnetic and mechanical properties which need to be investigated. It is further suggested that the magnetic anisotropy might be further increased by the incorporation of particles with shape anisotropy. It is known that rodlike iron oxide particles like goethite needles can easily be oriented by a magnetic field and are even able to form lyotropic phases (Lemaire et al. 2002); thus, such systems might be well suited to be incorporated into ferrogels and might even allow for the orientation of particles already during growth. Possibly even a similar synthetic approach involving a two-step mechanism with ripening from ferrihydride may be used (Burleson and Penn 2006). Another approach might be the fusion of spherical particles into rod-like particles on a micrometer length scale, as was already realized in solution (Vereda et al. 2013).

8.4.3 Conclusion

Ferrogels based on the natural biopolymer gelatin could be successfully synthesized. The degree of mineralization of these materials could be controlled by the number of repetition steps of the mineralization cycles. With variation of the synthesis method and the iron concentration, the particle size could be adjusted over one order of magnitude. The magnetic behavior of the materials could be controlled, yielding superparamagnetic and ferrimagnetic behavior. Ferrimagnetic particles possess a permanent magnetic moment, which manifested in the formation of macroscopic fibers in the ferrimagnetic gel. Drying in a magnetic field could thus create a high degree of order inside the ferrimagnetic ferrogels, supposedly leading to both anisotropic magnetic and mechanical properties.

The materials are biodegradable, and tuning of magnetic and mechanical properties makes them great candidates for applications in the biomedical field as well as for magnetic storage media.

8.5 Multifunctional Magnetic Composites

The chitin matrix from natural nacre can be used as a scaffold for the preparation of layered materials (see Fig. 8.11). Therefore the nacre is first sand blasted to remove the outer calcitic layer and can then be demineralized by various means via dissolution of the aragonite platelets by EDTA (Pereira-Mouriès et al. 2002) or dilute acetic acid (Siglreitmeier et al. 2015). SANS results showed how the structure of the nacre matrix survives the demineralization via acetic acid rather unchanged. Structures with a radius of gyration of R_g ~24.3 nm were observed, which are supposedly the holes in the matrix left over by the mineral bridges when taking into account the TEM images of embedded samples. With a diameter of approx. 68 nm, these are well suited to allow for the infiltration of gelatin. This was also proven by infiltration of coomassie-stained gelatin, which showed an even distribution of gelatin inside the matrix under the light microscope (Siglreitmeier et al. 2015).

In nacre, the scaffold chambers are filled with a silk-like hydrogel before mineralization (Levi-Kalisman et al. 2001; Nudelman et al. 2008). Inspired by this, a similar approach can be chosen, where gelatin can be infiltrated in the molten state

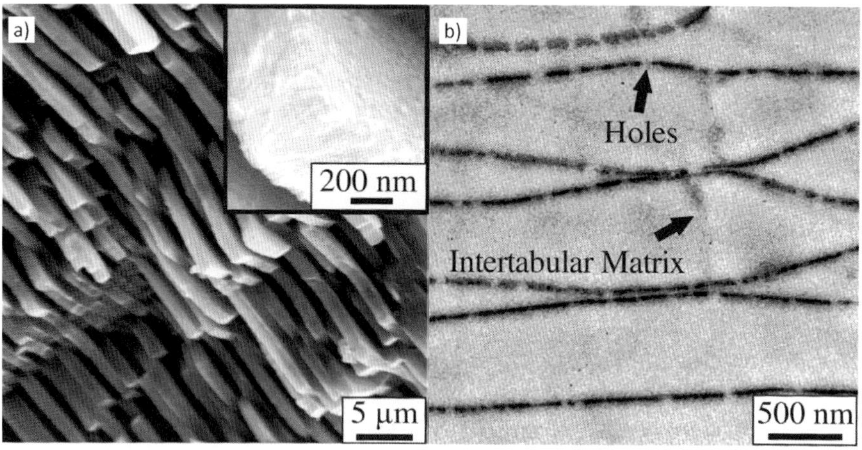

Fig. 8.11 (**a**) Layered structure of aragonite tablets as observed at a fresh fracturing surface of a *Haliotis laevigata* nacre shell and (**b**) microtome cut of a demineralized nacre matrix showing the compartmentalized structure with pores allowing for diffusion between the individual compartments (This figure was reproduced from Siglreitmeier et al. 2015)

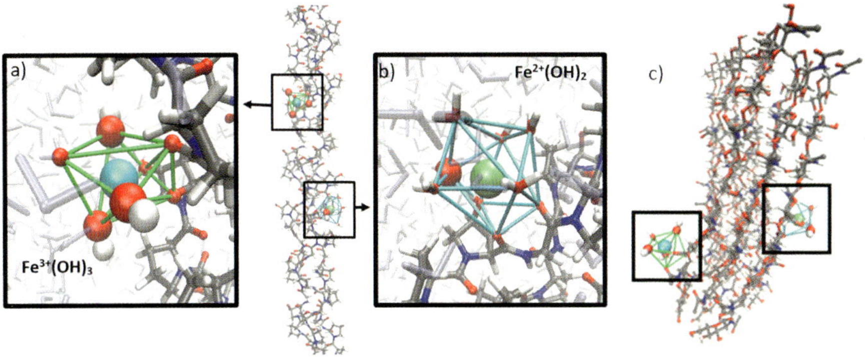

Fig. 8.12 (**a**, **b**) Triple helical structure of a collagen segment consisting of three (Gly-Hyp-Pro)$_n$ chains. (**a**) shows the association of a $Fe^{III}(OH)_3$ cluster to the collagen matrix, while (**b**) shows the coordination of a $Fe^{II}(OH)_2$ cluster. (**c**) shows an organic matrix resembling chitin, consisting of three layers of three poly-(1-4)-D-glucose chains (This figure was reproduced from Siglreitmeier et al. 2015)

via vacuum infiltration (Siglreitmeier et al. 2015). The mineralization with magnetite can then follow the established methods described before for the synthesis of ferrogels (Siglreitmeier 2015; Siglreitmeier et al. 2015). Iron salts are allowed to infiltrate the chitin-gelatin hybrid material and are then precipitated (and oxidized in case of the "partial oxidation" method) (Fig. 8.12).

Molecular simulations show how the presence of chitin may influence the magnetite mineralization compared to gelatin/collagen (Siglreitmeier et al. 2015). In the case of collagen, both ferric and ferrous hydroxide complexes would bind to the collagen triple helix. After relaxation in water, the iron (III) complex would bind stronger to the collagen than the iron (II) complex. The iron (II) complex was observed to partially dissociate in many cases. This also matches the observation that iron (III) chloride supposedly acts as a cross-linker between gelatin strains, notably increasing the stiffness of iron (III) loaded gelatin hydrogels. In contrast, both iron species form stable associations to chitin in simulations investigating the iron-chitin interaction. It was thus suggested that chitin can also act as a magnetite nucleator, and at least from a qualitative point of view, chitin might be slightly more favorable as a nucleation site than gelatin.

8.5.1 Superparamagnetic Layered Composite Materials

For composite materials mineralized via the coprecipitation method, a similar degree of mineralization as for the ferrogels can be achieved. Depending on the number of mineralization cycles, the content of inorganic material can vary between 15 and 65 wt.-%. As expected, microtome cuts of embedded materials show a constant increase of the magnetite particle density with an increasing number of mineralization cycles. In samples of mineralized chitin scaffolds without gelatin, it could

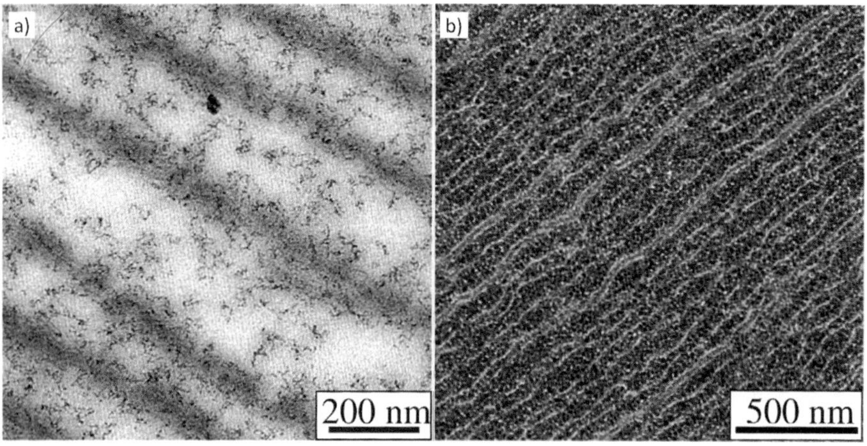

Fig. 8.13 TEM images of ultramicrotome cuts of the ferrogel prepared by the coprecipitation method after six reaction cycles in the swollen (**a**) and dried (**b**) state (This figure was reproduced from Siglreitmeier et al. 2015)

be shown that magnetite particles were only found on the chitin walls, supporting the suggestion of chitin also acting as a nucleator. In mineralized chitin-gelatin composite materials, the particles were found distributed over the scaffold cavities (Fig. 8.13). Particles on the chitin layers are slightly smaller than inside the gelatin hydrogel with approx. 8 compared to 12 nm. This is explained by the chitin acting as a heterogeneous nucleator, favoring the quick nucleation of smaller particles, shortening their growth time. Also, the interlayer distance is greater in samples containing gelatin, proving the role of the hydrogel not only as a nucleation site but also as a structural element in the pre-mineralization phase, keeping the cavities open, and later on as a nanoparticle carrier.

On freshly broken edges, the layered structure of these nacre-mimetic materials can be observed via SEM (Fig. 8.14a). Electron dispersive X-ray spectroscopy again shows how the iron is evenly distributed inside the scaffold chambers, while it is absent inside the organic scaffold layers.

Looking at SANS-USANS investigations of the layered hybrid material compared to the plain ferrogel, the chitin-gelatin-ferrogel shows a Q^{-1} instead of a Q^{-2} dependence at low Q values (<0.01 nm^{-1}, Fig. 8.14b). This suggests that the presence of the layered chitin scaffold supports the formation of a linear rather than branched arrangement of the magnetite nanoparticles.

Measuring the magnetic properties of the material via SQUID, similar values for the saturation magnetization as for the plain ferrogel are obtained with $M_s = 26$ emu/g at room temperature and 36 emu/g at 2 K. Typical superparamagnetic behavior is observed in both hysteresis loop and ZFC/FC measurements.

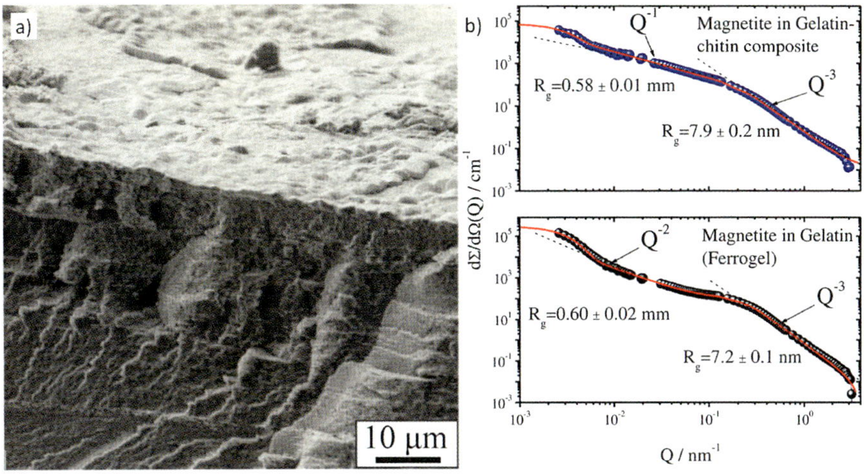

Fig. 8.14 (**a**) SEM image of breaking edge of a ferrogel composite, (**b**) SANS data comparing a ferrogel with a chitin-ferrogel composite (This figure was reproduced from Siglreitmeier et al. 2015)

8.5.2 Ferrimagnetic Layered Composite Materials

The synthesis of ferrimagnetic layered composite materials is analogous to the synthesis of superparamagnetic materials described before. Again, the nacre scaffold is demineralized and infiltrated with gelatin. Then the synthesis follows the procedure for ferrimagnetic ferrogels with infiltration of iron (II) sulfate, precipitation, and partial oxidation.

In the case of ferrimagnetic layered composites, only a mineralization degree of up to 45 wt-% compared to 65 wt-% for superparamagnetic materials can be achieved. Microtome cuts of the material in the swollen state show how the particles are mainly attached to the chitinous layers of the scaffold (Fig. 8.15a). This is in accordance with the simulation studies which suggested preferred binding of iron (II) to chitin over gelatin/collagen, leading to heterogeneous nucleation at the scaffold. Again, two different concentrations of iron (II) sulfate were used, leading to two different size regimes (50 and 90 nm), as well as particle morphologies comparable to the observations for the ferrimagnetic ferrogels described before. Compared to the superparamagnetic material, in the dried state, the particles are distributed less homogeneously between the chitin layers, and the packing is less dense, giving the material a more porous appearance in TEM images of microtome cuts (Fig. 8.15b). This is assumed to be due to the bigger particle size and higher polydispersity in the ferrimagnetic material, whereas the comparably monodisperse superparamagnetic particles assume a tight packing, which was also observed by SANS, as mentioned before.

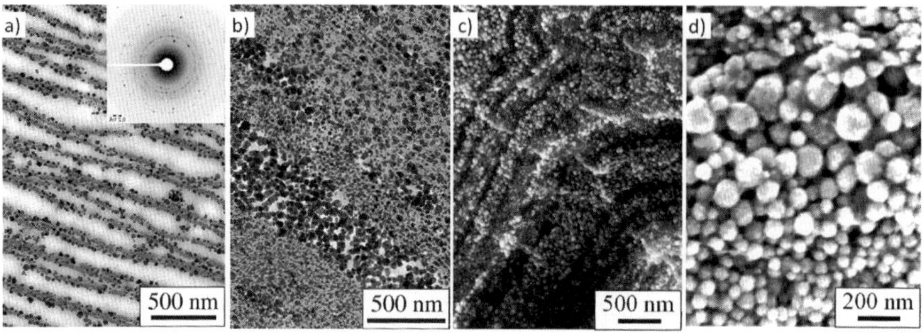

Fig. 8.15 TEM images of embedded ferrogel composites with an average particle size of 50 nm in the swollen (**a**) and dried (**b**) state, SEM image of dried composites with average particle sizes of 90 nm (**c**), and magnification of (**c**), (**d**)

The materials show ferrimagnetic behavior with hysteresis at room temperature and saturation magnetizations comparable to the values of the corresponding plain ferrogels.

8.5.3 Conclusion

The presented works showed the successful transfer of the ferrogel synthesis into the demineralized scaffold of nacre, leading to layered magnetic composite materials. The relevance of the gelatin for the mineralization process was shown. Similar as in the natural environment, the hydrogel not only influences the mineralization but also stabilizes the scaffold structure before and during mineralization. A degree of mineralization similar to the ferrogels can be achieved, and the material possesses a structure comparable to the chiton tooth composite surrounded by a nacre-like layered matrix.

8.6 Summary and Outlook

Magnetite nanoparticles can be synthesized at ambient conditions, and ways to control their size and morphology were shown based on iron concentrations and the precipitation routes. This control was further refined by the use of soluble gelatin as a cheap, easily obtainable, and biocompatible additive. With the coprecipitation method, approx. 12 nm-sized superparamagnetic particles can be created, while the partial oxidation method yields bigger particles ranging from 50 to 90 nm that show ferrimagnetic behavior. For the ferrimagnetic particles, flowerlike agglomerates of small subgrains were observed. These particles show single-crystalline SAED patterns, and their ferrimagnetism further confirms this single crystallinity. A HR-TEM study might be very interesting to prove this single crystallinity. Superparamagnetic and antiferromagnetic ferrogels can be created by applying these synthesis methods to the mineralization of gelatin hydrogels. The synthetic approach for ferrogels can be successfully transferred into demineralized scaffolds obtained from nacre demineralization, leading to hierarchically structured, layered magnetic composites.

The role of the gelatin in this material was shown to not only be a nucleation site for magnetite precipitation but also to stabilize the structure of the demineralized nacre during the mineralization process, in analogy to what is observed in the nacre biomineralization (Levi-Kalisman et al. 2001).

The magnetic properties of the nanoparticles were proven to be unchanged by incorporation into a ferrogel or a layered composite. The magnetic orientability of the ferrimagnetic nanoparticles could be used to align these particles, thus creating another level of hierarchy.

A main difference between the synthesized and the model materials is the limited degree of mineralization, which is as low as 65 wt.-% in the magnetic composites compared to, e.g., the mineral content of nacre with 95%.

Generally speaking, the proposed idea behind these composite materials was to take inspiration from three different materials, analyze their advantageous properties, select the structural features responsible for these properties, and combine them into one new material. In this case, a ferrogel mimicking the magnetite-organic composite of the outer shell of chiton teeth was chosen to give hardness to the material. The layered brick-mortar structure of nacre was applied to increase both fracture toughness and also stiffness and hardness of the material. Lastly, the high magnetic anisotropy of magnetosomes was copied with ferrimagnetic nanoparticles that allow the arrangement into linear structures.

The degree of mineralization might be further increased by an additional mineral phase, acting as a filler, that might be introduced after the magnetite mineralization. As the TEM images of microtome cuts of swollen gels show, these gels are not yet tightly packed and should still be accessible to additional diffusion-based mineralization steps. Minerals coming into question might be calcium phosphate, which is the natural "partner" of magnetite seen in the softer inner parts of chiton teeth (Saunders et al. 2009), calcium carbonate, which is the mineral seen in nacre (Heinemann et al. 2011), and silica, which can be found in combination with an iron oxide in the radular teeth of limpets, another mollusk comparable to chitons (Lowenstam 1971; Mann et al. 1986).

With an almost natural synthesis with many diffusion-based mineralization processes, the creation of gradient materials does also look promising. In many load-bearing biological materials, a core shell or gradient structure with a soft core and a hard shell can be observed (e.g., bone, teeth) (Bäuerlein 2008). A clever combination of different mineralization approaches or even making use of the self-healing properties of the ferrogel at low to medium degrees of mineralization will surely allow for the formation of such "multi-zone" or gradient materials.

To see if the basic idea behind these materials holds true, it would then be important to assess the mechanical properties of these materials via (nano-)indentation; abrasion, fracture, and bending tests; etc. Emphasis should also be laid on the possible anisotropy of the mechanical behavior, which is suggested by the layered structure, and on the additional influence that the orientation of the nanoparticles in a magnetic field might have. An additional stiffening and again a pronounced anisotropy in hardness and stiffness might be plausible.

Anisotropy can also be expected in the magnetic properties of the materials. Angle-dependent measurements with a magnetometer may show the macroscopic properties of oriented gels, while the nanoscopic magnetic interaction between individual particles might be observed by electron holography, a technique also applied to the particle chains of bacterial magnetosomes before (Simpson et al. 2005). The magnetic anisotropy might be further increased by the incorporation of morphologically anisotropic particles. The easily preparable goethite needles come to mind (Burleson and Penn 2006), which even have a natural counterpart in the goethite/silica composites found in the aforementioned limpet teeth (Barber et al. 2015).

Lastly, the only limiting factor regarding sample size and shape is the scaffold, which is obtained from natural nacre. Replacing the scaffold with an artificial chitin scaffold would abolish this limitation. There is already a lot of work done regarding creation of such layered biomimetic materials via freeze casting (Deville et al. 2006; Mao et al. 2016) and work regarding the combination of artificial scaffolds with the mineralization approach presented in the previous chapters is already ongoing. The continuing in-depth analysis of the material properties will show if they already represent a proof of concept of the central idea behind their design. Inspiration to further improve the overall material properties is existing and promising; thus, the properties of each of the model minerals may further be approached or even surpassed in the future.

Acknowledgments We gratefully acknowledge the financial support by the DFG priority program "Multifunctional Layered Magnetite Composites" (SPP1569). We want to thank Dirk Zahn and Tina Kollmann from the University of Erlangen for the simulation studies; Baohu Wu, Vitaliy Pipich, and Dietmar Schwahn from JCNS/MLZ for the scattering studies; and Damien Faivre from MPI Golm for support with the material design and early magnetite syntheses.

References

Ames W (1947) Heat degradation of gelatin. J Soc Chem Ind 66:279–284

Arakaki A, Webb J, Matsunaga T (2003) A novel protein tightly bound to bacterial magnetic particles in magnetospirillum magneticum strain AMB-1. J Biol Chem 278:8745–8750

Asenath-Smith E, Li H, Keene EC, Seh ZW, Estroff LA (2012) Crystal growth of calcium carbonate in hydrogels as a model of biomineralization. Adv Funct Mater 22:2891–2914

Barber AH, Lu D, Pugno NM (2015) Extreme strength observed in limpet teeth. J R Soc Interf 12

Bäuerlein E (2008) Handbook of biomineralization: biological aspects and structure formation. Wiley, Weinheim

Baumgartner J, Bertinetti L, Widdrat M, Hirt AM, Faivre D (2013a) Formation of magnetite nanoparticles at low temperature: from superparamagnetic to stable single domain particles. PLoS One 8:e57070

Baumgartner J, Dey A, Bomans PHH, Le Coadou C, Fratzl P, Sommerdijk NAJM, Faivre D (2013b) Nucleation and growth of magnetite from solution. Nat Mater 12:310–314
Brodsky PB, Werkmeister JA, Ramshaw JAM, (2004) Collagens and gelatins. In: Fahnestock S, Steinbüchel A (ed) Biopolymers: polyamides and complex proteinaceous materials II, vol 8. WILEY-VCH
Burleson DJ, Penn RL (2006) Two-step growth of goethite from ferrihydrite. Langmuir 22:402–409
Cornell RM, Schwertmann U (2003) The iron oxides: structure, properties, reactions, occurrences, and uses, 2nd edn. Wiley-VCH, Weinheim
Currey JD (1977) Mechanical-properties of mother of pearl in tension. Proc R Soc Ser B-Bio 196:443–436
Deville S, Saiz E, Nalla RK, Tomsia AP (2006) Freezing as a path to build complex composites. Science 311:515–518
Faivre D, Schüler D (2008) Magnetotactic bacteria and magnetosomes. Chem Rev 108:4875–4898
Goya GF, Berquó TS, Fonseca FC, Morales MP (2003) Static and dynamic magnetic properties of spherical magnetite nanoparticles. J Appl Phys 94:3520
Heinemann F, Launspach M, Gries K, Fritz M (2011) Gastropod nacre: structure, properties and growth—biological, chemical and physical basics. Biophys Chem 153:126–153
Heiss A, Jahnen-Dechent W, Endo H, Schwahn D (2007) Structural dynamics of a colloidal protein-mineral complex bestowing on calcium phosphate a high solubility in biological fluids. Biointerphases 2:16–20
Helminger M, Wu B, Kollmann T, Benke D, Schwahn D, Pipich V, Faivre D, Zahn D, Cölfen H (2014) Synthesis and characterization of gelatin-based magnetic hydrogels. Adv Funct Mater 24:3187–3196. excerpt figures shown are licensed under CC BY-NC-ND 3.0
Jolivet JP, Vayssieres L, Chaneac C, Tronc E (1996) Precipitation of spinel iron oxide: nanoparticle size control. MRS Online Proc Libr 432:145–150
Kang YS, Risbud S, Rabolt JF, Stroeve P (1996) Synthesis and characterization of nanometer-size Fe3O4 and γ-Fe2O3 particles. Chem Mater 8:2209–2211
Kniep R, Simon P (2007) Fluorapatite-gelatine-nanocomposites: self-organized morphogenesis, real structure and relations to natural hard materials In: Naka K (ed) Biomineralization I, vol 270. Springer, Berlin, pp 73–125
Lemaire BJ, Davidson P, Ferre J, Jamet JP, Panine P, Dozov I, Jolivet JP (2002) Outstanding magnetic properties of nematic suspensions of goethite (alpha-FeOOH) nanorods. Phys Rev Lett 88:125507
Levi-Kalisman Y, Falini G, Addadi L, Weiner S (2001) Structure of the nacreous organic matrix of a Bivalve mollusk shell examined in the hydrated state using cryo-TEM. J Struct Biol 135:8–17
Li H, Xin HL, Muller DA, Estroff LA (2009) Visualizing the 3D internal structure of calcite single crystals grown in agarose hydrogels. Science 326:1244–1247
Lowenstam HA (1971) Opal precipitation by marine gastropods (Mollusca). Science 171:487–490
Mann S, Perry CC, Webb J, Luke B, Williams RJP (1986) Structure, morphology, composition and organization of biogenic minerals in limpet teeth. Proc R Soc Lond Ser B Biol Sci 227:179–190
Mao LB, Gao HL, Yao HB, Liu L, Colfen H, Liu G, Chen SM, Li SK, Yan YX, Liu YY, Yu SH (2016) Synthetic nacre by predesigned matrix-directed mineralization. Science 354:107–110
Munch E, Launey ME, Alsem DH, Saiz E, Tomsia AP, Ritchie RO (2008) Tough, bio-inspired hybrid materials. Science 322:1516–1520
Niederberger M, Colfen H (2006) Oriented attachment and mesocrystals: non-classical crystallization mechanisms based on nanoparticle assembly. Phys Chem Chem Phys 8:3271–3287
Nudelman F, Shimoni E, Klein E, Rousseau M, Bourrat X, Lopez E, Addadi L, Weiner S (2008) Forming nacreous layer of the shells of the bivalves Atrina rigida and Pinctada margaritifera: an environmental- and cryo-scanning electron microscopy study. J Struct Biol 162:290–300
Özdemir Ö, Dunlop DJ, Moskowitz BM (1993) The effect of oxidation on the Verwey transition in magnetite. Geophys Res Lett 20:1671–1674

Penn RL, Banfield JF (1998) Imperfect oriented attachment: dislocation generation in defect-free nanocrystals. Science 281:969–971

Penn RL, Soltis JA (2014) Characterizing crystal growth by oriented aggregation. Cryst Eng Comm 16:1409–1418

Pereira-Mouriès L, Almeida M-J, Ribeiro C, Peduzzi J, Barthélemy M, Milet C, Lopez E (2002) Soluble silk-like organic matrix in the nacreous layer of the bivalve Pinctada maxima. Eur J Biochem 269:4994–5003

Pipich V, Balz M, Wolf SE, Tremel W, Schwahn D (2008) Nucleation and growth of CaCO3 mediated by the egg-white protein ovalbumin: a time-resolved in situ study using small-angle neutron scattering. J Am Chem Soc 130:6879–6892

Pouradier J, Venet AM (1952) The structure of gelatins. V. Degradation of gelatin in isoelectric solution. J Chim Phys Physicochim Biol 49:239–244

Reinhard Schrieber HG (2007) Gelatine handbook. WILEY-VCH, Weinheim

Roe RJ (2000) Methods of X-ray and neutron scattering in polymer science. Oxford University Press, New York

Saunders M, Kong C, Shaw JA, Macey DJ, Clode PL (2009) Characterization of biominerals in the radula teeth of the chiton, Acanthopleura hirtosa. J Struct Biol 167:55–61

Siglreitmeier M (2015) PhD thesis, University of Konstanz

Siglreitmeier M, Wu B, Kollmann T, Neubauer M, Nagy G, Schwahn D, Pipich V, Faivre D, Zahn D, Fery A, Cölfen H (2015) Multifunctional layered magnetic composites. Beilstein J Nanotechnol 6:134–148. excerpt figures shown are licensed under CC BY 2.0

Simpson ET, Kasama T, Pósfai M, Buseck PR, Harrison RJ, Dunin-Borkowski RE (2005) Magnetic induction mapping of magnetite chains in magnetotactic bacteria at room temperature and close to the Verwey transition using electron holography. J Phys Conf Ser 17:108

Sugimoto T, Matijević E (1980) Formation of uniform spherical magnetite particles by crystallization from ferrous hydroxide gels. J Colloid Interface Sci 74:227–243

Sun J, Zhou S, Hou P, Yang Y, Weng J, Li X, Li M (2007) Synthesis and characterization of biocompatible Fe3O4 nanoparticles. J Biomed Mater Res A 80A:333–341

Taylor RM, Barbara, Self P (1987) Magnetite in soils; I, The synthesis of single-domain and superparamagnetic magnetite. Clay Miner 22:411–422

Teja AS, Koh P-Y (2009) Synthesis, properties, and applications of magnetic iron oxide nanoparticles. Prog Cryst Growth Charact Mater 55:22–45

Tewari PH, McLean AW (1972) Temperature dependence of point of zero charge of alumina and magnetite. J Colloid Interface Sci 40:267–272

Tlatlik H, Simon P, Kawska A, Zahn D, Kniep R (2006) Biomimetic fluorapatite–gelatine nanocomposites: pre-structuring of gelatine matrices by ion impregnation and its effect on form development. Angew Chem Int Ed 45:1905–1910

Vayssières L, Chanéac C, Tronc E, Jolivet JP (1998) Size tailoring of magnetite particles formed by aqueous precipitation: an example of thermodynamic stability of nanometric oxide particles. J Colloid Interface Sci 205:205–212

Vereda F, de Vicente J, Morales MDP, Rull F, Hidalgo-Alvarez R (2008) Oxidation of ferrous hydroxides with nitrate: a versatile method for the preparation of magnetic colloidal particles. J Phys Chem C 112:5843–5849

Vereda F, de Vicente J, Hidalgo-Alvarez R (2013) Synthesis and characterization of single-domain monocrystalline magnetite particles by oxidative aging of Fe(OH)2. J Colloid Interface Sci 392:50–56

Vergés MA, Costo R, Roca AG, Marco JF, Goya GF, Serna CJ, Morales MP (2008) Uniform and water stable magnetite nanoparticles with diameters around the monodomain–multidomain limit. J Phys D Appl Phys 41:134003

Walz F (2002) The Verwey transition – a topical review. J Phys Condens Matter 14:R285

Wandrey C, Bartkowiak A, Harding SE (2010) Materials for Encapsulation. In: Zuidam NJ, Nedovic V (eds) Encapsulation technologies for active food ingredients and food processing. Springer New York, New York, pp 31–100

Weaver JC, Wang QQ, Miserez A, Tantuccio A, Stromberg R, Bozhilov KN, Maxwell P, Nay R, Heier ST, DiMasi E, Kisailus D (2010) Analysis of an ultra hard magnetic biomineral in chiton radular teeth. Mater Today 13:42–52

Wu B (2015) Bio-inspired Magnetite Mineralization in Gelatin Hydrogels: A Small Angle Scattering Investigation. PhD thesis, Universität Konstanz

Xu AW, Ma YR, Colfen H (2007) Biomimetic mineralization. J Mater Chem 17:415–449

Printed by Printforce, the Netherlands